MW00978409

GLOBAL IMPACT

GLOBAL IMPACT

The New Telecommunication Technologies

Loy A. Singleton

Harper & Row, Publishers, New York
BALLINGER DIVISION
Grand Rapids, Philadelphia, St. Louis, San Francisco
London, Singapore, Sydney, Tokyo

International Standard Book Number: 0-88730-259-9

Library of Congress Catalog Card Number: 89-6559

Printed in the United States of America

Library of Congress Cataloging in Publication Data

Singleton, Loy A.
Global impact : the new telecommunication technologies / by Loy A. Singleton.
p. cm.
Includes index.
ISBN 0-88730-259-9
1. Telecommunication systems. I. Title.
TK5101.S534 1989
384—dc20 89-6559
CIP

89 90 91 92 HC 9 8 7 6 5 4 3 2 1

To Sarah, Christine, and Sandy

CONTENTS

PART VIII
Conclusion

PREFACE

This book was written with two goals in mind. The first was to introduce a number of the new electronic media, explaining their origins and operation in a nontechnical manner. The other goal was to do this in an international context or at least to provide international examples of the application and development of these technologies. To the extent that these two aims have been achieved, readers should reach the end of the book not only with a basic understanding of each technology or medium, but also with the perception that the new electronic media are in many ways international phenomena with global implications.

The material that follows is organized into eight parts. The three introductory chapters deal with the general characteristics, basic technical concepts, and regulatory context of the new electronic media. They are intended as a brief introduction for the uninitiated or a quick refresher course for those revisiting this subject.

Each of Parts 2 through 7 (Chapters 4 through 20) encompasses a major development or trend of the last twenty years, such as the rise of satellite communications, the evolution of the home TV set from a broadcast signal receiver into a video information and entertainment appliance, the transformation of the cable and telephone industries and their potentially competitive roles, the development of electronic publishing and video retailing services, the increasing use of telecommunication technologies as substitutes for transportation, and the growing ability of telecommunication technology to reach any person anywhere. Chapter 21 offers some conclusions about the global, paradoxical nature of the new electronic media.

Readers familiar with the predecessor to this book, *Telecommunications in the Information Age,* will find that although this is an entirely new treatment of these topics, the approach to the material once again emphasizes practical concerns. As with the previous book, considerable space is devoted to the history of each technology or medium. The story of the origin and growth of these new technologies is not only inherently interesting, it is instructive, often revealing underlying similarities and patterns in the development of different systems.

Some readers will notice the chapter subdivisions ("Background," "How it Works," "Applications," and "Forecast") used in the previous book have been replaced by topical headings specific to each chapter. This reflects an effort to organize each chapter in a way that tells the story of that particular industry or technology most effectively. Nevertheless, the progression in most chapters is similar, starting with background and including basic technical aspects, current examples, and industry issues.

As anyone who has undertaken such a project can testify, the final draft of a book always ends up being a compromise between the ideal in mind at the outset and inevitable limitations on time, energy, and resources. These contents represent an effort to gather and include current information interesting and relevant to readers with limited knowledge of these subjects. In doing this I have tried to keep in mind what seems meaningful to the students in my own "new technologies" classes and the kinds of questions I am asked frequently by colleagues from other fields. Readers should keep in mind that dramatic changes can take place quite rapidly and even unexpectedly in the realm of new electronic media. Although this material was current going to press, recent issues of industry periodicals cited throughout the book should be consulted for developments since publication.

Many people have contributed to this book in a variety of ways. I am especially grateful to the managers, entrepreneurs, regulators, and colleagues who took time to answer questions, send information, and offer advice and encouragement. Among them were Barbara Moran, Robert Whitehill, and Charles Gratch, Federal Communications Commission; Jackie Biel, Kompas-Biel Associates; Cara Hancock, National Satellite Paging; Dr. Byron St. Clair, TTC; John Reilly, Jefferson Communications; Curt Bradley, Wireless Cable Association; Jim Clark, Conifer Corporation; David Daugherty, Frank Magid Associates; Gil Gordon, Gil Gordon Associates; Nancy Knauer, Telocator of America; Neil Suhre, Advanced Voice Systems and Donnelly Directories; Mary Dale Walters, Telaction Corporation; Joseph Martin, Carolina Telephone; Prof. Jorge Reina Schement, School of Library and Information Science, Rutgers University; Prof. John Freeman, RTF Department, Texas Christian University; and Prof. Everett Rogers, Annenberg School of Communications, University of Southern California.

In the Department of Radio, Television, and Motion Pictures, University of North Carolina at Chapel Hill, Prof. John Bittner again provided invaluable support and expert advice, drawing on his extensive writing and research experience. Chairman Gorham Kindem's accommodating class scheduling made time available for research and writing. Ian Gibson provided timely research assistance. Professors Richard Elam, Seth Finn, and Anne Wadsworth contributed helpful articles, information, and ideas. Lastly, my appreciation to the students in my "new technologies" classes, undergraduate and graduate, who helped shape this book through their classroom comments, questions, and enthusiasm.

PART I

An Overview of Telecommunications

1

Perspectives on Electronic Communication

Today's new electronic media make possible things reminiscent of yesterday's science fiction: an entire encyclopedia stored on a four-inch disk; shopping at home by television; feature films on palm-size cassettes; talking computers that take your phone message and then page you a continent away; international videoconferencing; dozens of channels of TV programs, including round-the-clock news and weather and live coverage of Congress; pocket-size wireless telephones; backyard satellite dishes; and so forth. Remarkably, none of these systems and services were available as little as twenty years ago; some are products of the last decade.

A global perspective is quite important to anyone seeking to understand these new electronic media today. In past decades Americans tended to take a somewhat chauvinistic view of telecommunication technology. After all, the telegraph and telephone were essentially U.S. inventions, and many of the basic advances in telecommunication technology regularly had come out of U.S. laboratories, among them the transistor, the laser, microelectronics, communication satellites, the first videotape recorders, and so forth. Furthermore, AT&T was the world's largest telecommunications company, and U.S. consumer and industrial electronics helped set standards worldwide.

Today, however, that familiar situation has rapidly given way to a new and more competitive international environment. AT&T, for example, is no longer the global giant it once was, having been split up and downsized by deregulation. Imported telephones and other telecommunication equipment are now common in U.S. homes and offices. Although as recently as 1981 the United States enjoyed an $817 million trade surplus in telephone and telegraph equipment, by 1986 the surplus had become a $2.5 billion deficit.[1]

In most stores American-made consumer electronics have been largely replaced by imported televisions, radios, and VCRs. The state-of-the-art in video, for example, is predominantly the product of Japanese or European—not American—manufacturing. In the mid-1970s there were fifteen U.S.-owned color television manufacturers; today only Zenith remains, and the others are sold to foreign companies or out of business.[2]

Perhaps symbolic of the general trend, RCA's consumer electronics division, once a world standard bearer, was bought by General Electric in 1985, then sold with GE's own consumer electronics interests to Thomson, a French electronics company, in 1987. Through RCA, Thomson now controls a larger share of the U.S. market than any U.S.-owned firm. Entering the 1990s the world's consumer electronics leaders along with Thomson are N.V. Philips of the Netherlands and Matsushita of Japan.[3]

In short, many former assumptions about the U.S. role in the present and future development of electronic media are being challenged by new realities. It is hoped that in addition to providing a painless introduction to the "how" of the new technologies, this book outlines some of these new international realities.

Considering that we can pick up a telephone and place an international call or turn on a television set and see the latest news from Washington via satellite, it is easy to forget the awe inspired by the world's first new electronic medium. Although humans had been able to "telecommunicate"—communicate at a distance—for as long as there had been signal fires, mirrors, or runners, Morse's telegraph allowed instantaneous communication over great distances, further than the eye could see, even at night or in bad weather. It had economic and social implications that seemed truly worthy of the melodramatic question Morse transmitted at the historic first telegraph demonstration in 1844: "What hath God wrought?"

Today we wonder not as much at the mere existence of instantaneous electronic communication as at its remarkable variety and technological complexity. Industry observers, consumers, and even communication professionals sometimes are bewildered by the expanding variety of new electronic media. With new systems and services continually appearing, it has become a challenge simply to keep abreast of the latest developments. Chapters 4 through 21 survey many of these developments, but Chapters 1, 2, and 3 first will take a broader view, discussing common characteristics of electronic media, the media's technological hierarchy, and the paths of electronic communication.

Some Attributes of Electronic Media

Consider two electronic media as apparently different as broadcast television and electronic voice messaging. How can we readily summarize their differences or perhaps describe any similarities? Such comparisons are necessary every day in media regulation, marketing, and research and are often made using widely differing criteria. Several general attributes or criteria can be used to describe or compare the electronic media discussed in the chapters that follow.[4]

Mobility. Is the medium normally used by senders and receivers in set locations (fixed service), or may either one be used in a vehicle, either land, sea, or air (mobile)? A major trend in electronic media in recent years has been the development of mobile versions of what were traditionally fixed services, as with telephone and television.

Transmission Format. Today's electronic media convey their messages in four basic formats: image, text, audio, and data. Although some media are associated with one format exclusively, others convey more than one type of content. Television carries images, audio, and text, for example. Residential phones most commonly transmit only voice or data.

Transmission Capacity. Messages vary in complexity within and across the format categories. Beethoven's Fifth Symphony is a much more complex audio signal than a recorded phone message. Moving color video is more complex than still-frame monochrome. Generally, the greater the complexity of the message, the greater must be the transmission speed and processing capability—the transmission capacity—of the electronic medium. For example, Morse's telegraph system could transmit only a few characters per second; a modern television transmitter sends enough information every thirtieth of a second to reconstruct the color and location of every dot in the 525 lines that make up a single frame of video. This is also why a satellite can carry twelve FM radio stations or perhaps 1,200 telephone conversations in the same channel required by a single color television signal. Capacity can be thought of in terms of a spectrum, ranging from low-capacity transmission to high. These are, of course, relative terms. As telecommunication technology continues to evolve, yesterday's high-capacity system rapidly becomes today's low-capacity system.

Immediacy. A real-time system allows communication to take place instantaneously; a delayed system does not. Real-time communication is not always a virtue, nor is delay necessarily a vice; it depends on the situation. The telephone system, for example, and live television transmission permit conversation or attendance at events over great distances. But they also require your real-time presence, which limits your ability to use them. As technology developed that permitted the storage and recall of such transmissions, telecommunication could take place over time as well as distance. VCRs and audio- and videodisk players are store and playback devices that enable us to see and hear certain types of transmissions at our convenience. A telephone answering machine allows us to transform a real-time system, the telephone, into a delayed system when such delay is advantageous.

Sender/Receiver Combination. Electronic media are often described in terms of this characteristic—that is, according to the combination or pattern of senders and receivers they link together. For example, the term *mass media* is used to describe radio, television, and other telecommunication systems when they link a single sender to a multitude of receivers.

In very general terms, we can say that electronic communication may link three broad categories of senders and receivers: persons, groups of persons (meaning anything from a small group to millions), and machines (meaning computers or other devices programmed to transmit or receive information). These three categories suggest nine sender–receiver combinations, as Table 1–1 illustrates.

Table 1–1. Telecommunication Sender–Receiver Combinations

Person to Person	Person to Group	Person to Machine
Group to Person	Group to Group	Group to Machine
Machine to Person	Machine to Group	Machine to Machine

The systems and technologies described in subsequent chapters permit many of these sender–receiver combinations. Videoconferencing, for example, can take the form of person–person or person–group communication. Voice messaging systems involve person–machine interaction, as does using a videotext service or an electronic shopping catalogue. Although beyond the scope of this book, machine–machine communication is becoming common in factories as computers monitor and control robotic devices on assembly lines.

Range or Coverage. Broadcasting media are commonly categorized according to the range of their signals: local, regional, national, and so forth. Other types of electronic media also have differing coverages. A cable system may cover a single condominium complex or an entire city; a paging system may reach to the city limits or be extended to encompass an entire state or region; a teletext system may reach only a local audience or a national audience via satellite distribution. Coverage can be increased in two basic ways: by directly increasing the signal strength or transmission lines of an existing system or by linking less powerful systems into networks.

Interactivity. Electronic communication has a kind of directionality, and media are often categorized accordingly. In a one-way medium the transmission is all in one direction, and no real-time interactivity can take place. In a two-way system transmission and reception take place freely in both directions, permitting a high level of interactivity. A telephone system is two-way; interaction between sender and receiver is immediate and spontaneous. On the other hand, an FM radio station is a one-way telecommunication system. It can be made two-way only by supplementing it with the telephone system, as in a call-in show.

Communication Technology, Systems, and Networks

Technology is the aspect of electronic communication that comes to mind most readily, the physical "stuff" that must be hung on poles, bolted together, plugged into the wall, or held in the hand. This technology can be as simple as Morse's first telegraph or as complex as an international computer network. It includes all the physical devices—the hardware—that must be in place for electronic communication to occur. Unfortunately, those of us who are not engineers tend to use ambiguous language when talking about the many forms telecommunication hardware can take. For example, we tend to use terms like *new technologies* and *new electronic media* interchangeably. One way of bringing some order to these labels is to think of the technology as constituting a kind of

Figure 1–1. The Telecommunication Technology Pyramid

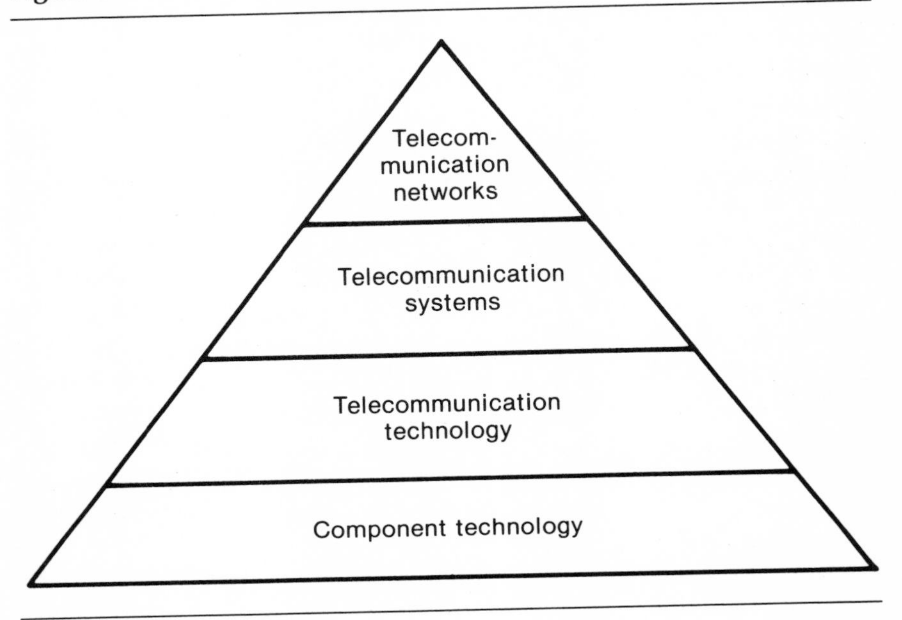

hierarchy or pyramid. This technology pyramid has four levels, and each level is of increasing complexity: component technologies, communication technologies, systems, and networks.

As Figure 1–1 suggests, the foundation of the pyramid is component technology. Component technologies must be developed or invented and then interconnected or bundled together to become a recognizable communication technology. Component technologies do not necessarily look like communication technologies by themselves. A battery, a piece of wire, a transistor, even a laser, for example, all have a variety of applications besides communication.

For instance, the first electronic medium, Morse's telegraph, was a bundle of several component technologies, including copper wire, batteries, and the separate parts of the telegraph key itself. Individually none of these components could be called a communication technology. Assembled in a unique way, these components became the telegraph instrument we see in textbooks and museums, recognizable as a device used for electronic communication. Communication technology, then, is the second level of the pyramid.

Most of the time specific electronic communication technologies must be interconnected in some way to provide the service we associate with them. Morse's telegraph key had to be linked to another key and a method of encoding information developed before any communication could take place. Morse's code, which made high-speed telegraphy possible, provided the missing technological link in the chain of technologies comprising the telegraph.

Morse's contribution obviously went beyond inventing or designing a particular communication technology. He brought together technologies to create a working telegraph system. When technologies are linked together in a manner that permits communication, they make up a telecommunication system, the middle level of our technology pyramid. Systems are the form of technology that most of us actually use in day-to-day life and usually refer to as electronic media.

A contemporary example is a local cable system. It consists of many component technologies assembled into communication technologies (like satellite receiving dishes, coaxial cable, line amplifiers, converter boxes, television receivers, and so on). These in turn have been linked and configured in a way that allows distribution of television programming to subscribers from many sources—a cable *system.* A local radio or television broadcasting station, telephone company, cellular mobile radio service, videotext service, and local radio paging service are all examples of technology at the system level, even though we refer to them as stations, services, media, or whatever. A system, then, is the least combination of technology that can provide electronic communication service or functions.

But individual systems alone usually do not provide communication on the national or mass scale that we now take for granted. Large-scale or mass communication usually is made possible by linking systems. Interconnected systems make up networks, the fourth level of the technological pyramid. For example, modern long-distance telephone service is made possible by linking local telephone systems using microwave and satellite technology and computer-controlled switching and billing systems. Likewise, a national cable network is made possible through the interconnection of local cable systems by satellite, as is a television or radio broadcasting network.

Much of the history of electronic media is the evolution of technologies into systems and systems into networks. Bell's first working phone soon became a system, the country's first local telephone exchange. That system and other local exchanges were soon interconnected to become an intrastate and then interstate network. A similar evolution took place in the 1920s with radio, the 1940s with television, the 1970s with cable, and the 1980s with paging and cellular phone systems. The chapters that follow attempt to maintain this distinction between technology, systems, and networks.

Paths of Electronic Communication

Electronic communication can be accomplished in a variety of ways, depending on the desires and technical capabilities of the sender and receiver. Figure 1–2 suggests the series of steps or various paths that information follows in the process of electronic communication. The process begins with information (audio, text, image, or data) in a natural form—a form that can be perceived through sight or sound.

This information must be transformed or encoded into a form that makes electronic transmission possible. For example, the voice of a radio news reporter, a person speaking into a telephone, and the image of a TV weather map all contain information that we detect with our ears as patterns of air vibration or with our eyes as patterns of light when we are in the same place as they. These "natural" information signals must be converted into electronic information signals and changed into a form that cannot be perceived directly by the senses. This conversion or encoding (step A in Figure 1–2) is accomplished by the reporter's microphone, the telephone's mouthpiece, the TV camera's video pickup tube, and so forth.

Once converted to electronic form, the information signal may be transmitted immediately (step B). Transmission can be accomplished either over the air or by wire. Examples of immediate (real-time) transmission include a live television broadcast, a telephone conversation, or a CB radio transmission. Reception (C) is accomplished by a telephone handset, television receiver, or perhaps a car radio, which then use loudspeakers and a picture tube to decode the information (D)—that is, to change it from electronic form to a natural signal (sight and sound) once again. This path (A–B–C–D) is the original form of telecommunication: live transmission of information with real-time (instantaneous) reception and decoding. Morse's telegraph, Marconi's wireless, contemporary live television, and a phone call across the street are all examples of this same basic process; it's essentially the same whether transmission takes place through the air or through a wire.

The development of electronic information storage technology opened up a number of other paths for electronic information transfer, as Figure 1–2 suggests. Electronic information storage can be accomplished in a number of ways, including electromagnetic media (audio- and videotape,

Figure 1–2. Paths of Electronic Communication

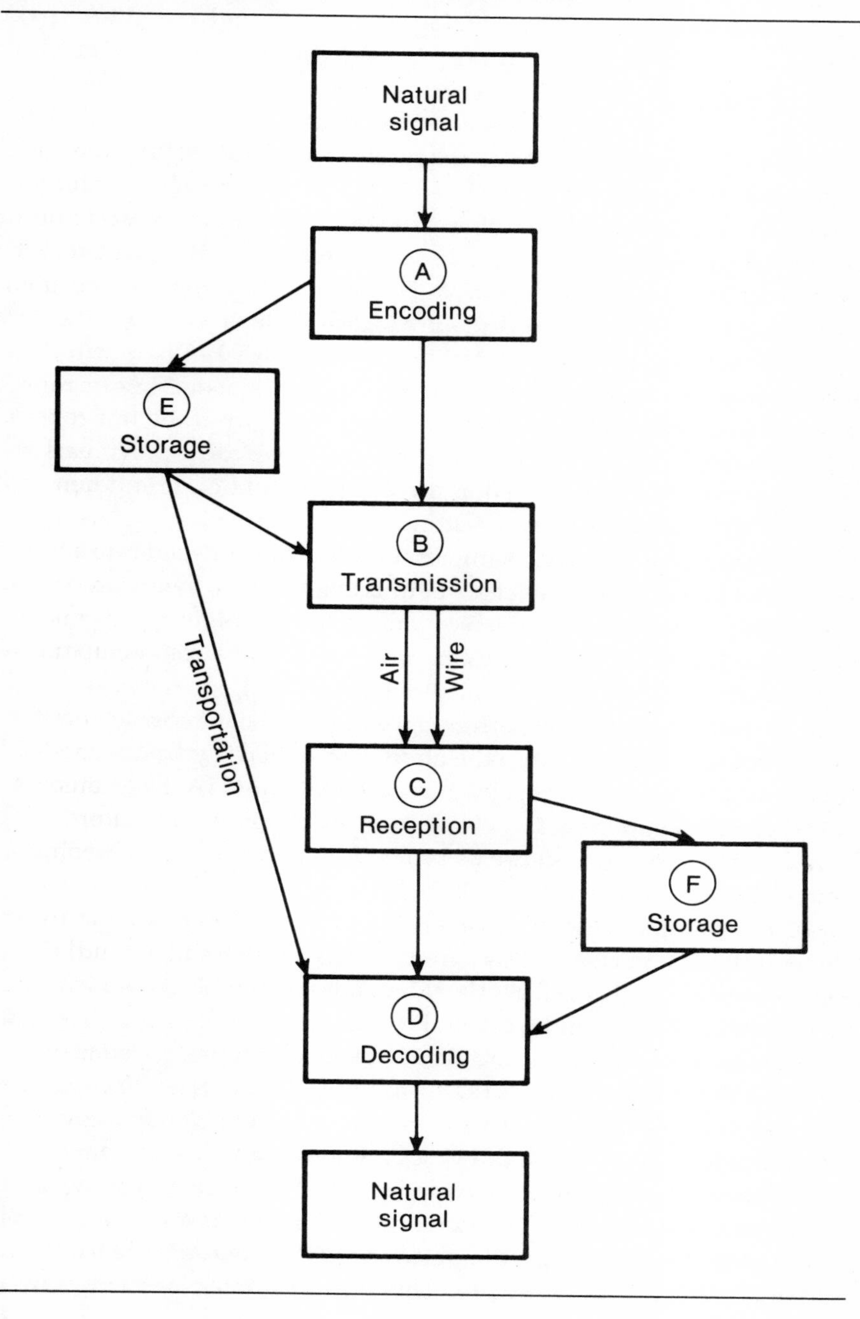

computer disks), computer chips, optical media (audio- and videodisk), and conventional audiodisks. Storage detours (E and F) can take place before transmission, in which case they are controlled by the sender, or after reception, when they are controlled by the receiver.

After being stored (E) for seconds or decades, information may follow either of two major paths to the receiver. On the first path it is transmitted and received by air or wire (E–B–C), after which it is decoded immediately (D) or takes another storage detour (F) to await decoding later. For example, any television broadcast that originates on tape rather than live follows the first path (A–E–B–C–D). If, however, the home viewer uses a VCR to record the broadcast for later viewing, the path includes a detour to F before decoding (A–E–B–C–F–D).

On the second major path after storage, the encoded information is not electronically transmitted at all; rather, it is physically transported. This is the path (A–E–D) followed by a compact disk purchased at a record shop, a cassette rented at a video rental outlet, or a floppy disk bought at a computer software store.

It is worth noting that as information flows from sender to receiver in Figure 1–2 it must have legal "permission" to pass every point along its way or it is considered stolen or pirated. For example, cable programming finding its way to a person via a stolen cable converter box has failed to pass point C legally, breaking federal and, in many cases, state law. Or a rock star's stage performance that has been recorded on a cassette recorder smuggled into the auditorium, then duplicated, and sold underground has been illegally encoded and stored (A–E), then passed to D, violating copyright law. Finally, a pirate radio station violates FCC rules and federal law because its programming passes point B without a license to transmit.

It is also interesting to consider that no one information pathway is inherently superior to another. They all exist as alternatives, and the selection of one path over another depends on the objectives and constraints of the sender and receiver. For example, instantaneous live transmission of information has the virtue of speed, but that speed is a virtue only if the information is in the form desired. Any television news reporter who has lost control of an eyewitness interview while covering a story live can testify to the drawbacks of instantaneous telecommunication. Likewise, the former shah of Iran was overthrown by a revolution in which a delayed form of telecommunication played an important role: the humble audio cassette smuggled in, copied, and passed from hand to hand, was effectively used to maintain communication links between revolutionary leaders outside the country and followers within when government forces tightened control of broadcasting and wire communications systems.

Electronic communication is a process through which information is changed into electronic form for transmission to one or more receivers. Electronic transmission can take place instantaneously, or it can be delayed using electronic storage devices that make it possible to overcome both distance and time as impediments to human communication.

The technology of electronic communication can be thought of as a hierarchy of increasing complexity, helping to differentiate between technological components, networks, and systems. Finally, all forms of electronic communication have common attributes that are used to compare them and that may be useful in understanding and contrasting the media surveyed in the following chapters.

Notes

1. National Telecommunications and Information Administration, "Telecom 2000: Charting the Course for a New Century" (Washington, D.C.: U.S. Government Printing Office, 1988), 5.
2. "Zenith, after GE's Exit, Is Last Big TV Maker in U.S.," *Wall Street Journal*, 23 July 1987, p. 23.
3. Laura Landro and Douglas R. Sease, "General Electric to Sell Consumer Electronics Lines to Thomson SA for Its Medial Gear Business, Cash," *Wall Street Journal*, 23 July 1987, p. 3.
4. See David Cleevely and Richard Cawdell, "A Telecommunications Taxonomy, *Telecommunications Policy* (June 1986): 107–19, for a fully developed taxonomy that exceeds the scope of this chapter and from which some terms were adapted for this discussion.

2

Some Basic Technical Concepts

More often than not, discussions of new electronic media at one point or another include terms like *pulse-code modulation, bandwidth, gigahertz,* or *spectrum management.* This chapter briefly reviews these and other basic concepts that underlie much of the material discussed in subsequent chapters. It focuses on two fundamental areas: the transmission process and the nature and management of the electromagnetic spectrum.

Fundamentals of Electronic Transmission

Electromagnetic Energy. We are all familiar with the electrical power system at home and elsewhere. Recall that the flow of electricity in a wire is caused by the movement of electrons. The form or characteristics of this flow or current can be manipulated in a number of ways. For example, electrical energy can be made to flow in a single direction through a wire (direct current or DC) or can be caused to alternate direction (alternating current or AC). Alternating currents have certain measurable characteristics, including frequency, speed, and wavelength.

Frequency. The frequency of an alternating current is the number of alternations, called cycles, it completes in a second. For example, household electrical current in the United States has a frequency of sixty cycles per second, referred to as 60 Hertz; 1 Hertz equals one cycle per second. The electromagnetic energy used for transmission by the technologies discussed in this book alternates or cycles at extremely high rates, and a metric reference system is used to refer to their frequencies. A 50 kilohertz wave alternates at a frequency of 50,000 cycles per second and is written "50 kHz." A 50 megahertz wave completes 50 million cycles per second (50 mHz); and a 50 gigahertz wave, 50 billion cycles per second (50 gHz).

Speed. All electromagnetic energy flows at the same speed, the speed of light: 186,000 miles per second. This is true whether the energy travels through air, water, or solid objects. The energy may be scattered or absorbed by the medium through which it travels, but its transmission speed remains the same. For most forms of electronic communication this speed is virtually instantaneous. However, as fast as it is, light speed can be noticeably slow in some newer forms of electronic communication. For example, it takes about one-third of a second for a signal to travel from the sender to the satellite to the receiver in a satellite-relayed telephone call, a trip of at least 44,600 miles.

Wavelength. Because all electromagnetic energy travels at the same speed, the physical length of any single wave is related to the wave's frequency. For example, a 1 Hertz wave and a 10 Hertz wave both always will travel the same distance in a second because they are both traveling at light speed. The 1 Hertz wave completes one alteration or cycle during that one-second trip. Its wavelength, then, is 186,000 miles. The 10 Hertz wave completes ten cycles during the one-second trip. Each of its cycles, therefore, can be only one-tenth as long: 18,600 feet. Thus the lower the frequency, the longer the wavelength, and the higher the frequency, the shorter the wavelength.

Carrier Waves. When electromagnetic energy is used to transport information, it is called a *carrier wave* or *carrier.* In AM, FM, and TV broadcasting the frequency you tune to with your radio or TV receiver is the carrier wave frequency. First, images, sound, or data are converted by a microphone, camera, or keyboard into electronic form—that is, into an electronic information signal. Then that signal is impressed on the carrier wave, a process called modulation.

Modulation. Currently three types of modulation are commonly used in electronic communication. Frequency modulation (FM) continuously shifts the frequency of the carrier wave up or down from its original frequency, while the other characteristics of the wave remain constant. As an analogy, suppose that the beam of light from a flashlight was to be used as a kind of visible carrier wave. First we would decide on a frequency for our carrier; that is, we would select a color for the light beam.

To frequency modulate the light beam we would continuously shift its color (frequency) slightly toward the higher or lower end of the visible spectrum while the beam remained on. In FM radio broadcasting the carrier wave frequency—the frequency you tune to—is shifted to somewhat higher and lower values, re-creating the variations in the original information signal.

Amplitude modulation (AM) continuously varies the strength of the carrier wave. In the flashlight analogy, this would mean that the color of the beam (frequency) remains the same, but the brightness increases and decreases slightly in a pattern that, once again, reproduces the changing values of the original information signal.

A third technique, pulse code modulation (PCM), does not continuously modulate the carrier as the information signal changes. First, rather than continuously reproducing the electrical pattern generated by the information signal, it samples the information signal thousands of times per second. Each of those measurements then is rounded off and its numeric value converted to binary form—the on/off code of computer data. This digital computer code is then impressed on the carrier. Rather than resembling a complex, continuously varying wave, the modulated carrier resembles a simple series of pulses, much like beads on a string. This is digital transmission, a technique that is the hallmark of many new electronic media. In the flashlight analogy, the beam would remain the same color and brightness but flash off and on rapidly, resembling a naval signal light.

Transmission by Wire. A modulated carrier can transport information either through a wire or through the air. The earliest electronic communication systems, the telegraph and telephone, were wireline systems. The information-carrying capacity of a wireline system depends on the type of wire or cable used. Most basic is the open-wire line, simply a bare wire hung on poles, capable of handling perhaps a dozen telephone signals simultaneously. Cables—copper wires insulated with metallic and plastic sheathing—have a greater information transmission capacity. For example, neighborhood telephone lines often carry twenty-four signals on each pair of twisted copper wires in a multiwire cable. Lastly, there is coaxial cable, which can be used to transmit several hundred telephone signals, or perhaps several dozen television channels, in a single line. Coaxial cable of about an inch in diameter is commonly used for signal distribution by local cable television systems.

Over-the-Air Transmission. A carrier wave used to transmit information through the air is called a radio wave. Transmission through the air is possible because electromagnetic energy flow does not necessarily stop at the surface or edge of a wire but can continue into space, radiating out in many directions. When an AC current "escapes" into space, the energy travels or radiates in the form of a radio wave. The harnessing of radio waves toward the end of the nineteenth century provided a means of wireless radio communication with greater mobility and range than wireline media. Today most electronic media combine wireless and wireline methods to transmit information. For example, a long-distance telephone call starts out in a wire but then may be related by satellite or radioed to an automobile; a television station originally transmits its programs through the air, but those programs may find their way to viewers over a local cable system wire.

Transmitters and Receivers. In wireless transmission systems, radio waves are injected into the atmosphere by a transmitter. A transmitter generates a carrier wave of a particular frequency, modulates it with the information signal, amplifies it, and radiates it into the environment using a transmitting antenna. When the radio wave strikes a receiving antenna, its presence is detected by the set of components comprising the receiver. The receiver selects the appropriate carrier wave frequency, amplifies it, and removes the information from the carrier.

Antennas. An antenna is a length of metal conductor that allows transmitters and receivers to emit or capture airborn carrier waves. Antennas are most efficient when they are the same length as the wavelength of the signal they are sending or receiving. However, in many communication applications this is not practical. For example, the 550 kHz signal of a radio station transmitting at 550 on the AM dial has a wavelength of about 1,750 feet. This means the ideal transmitting antenna is 1,750 feet long, requiring a much taller tower than is practical for most such applications.

The solution lies in antennas that are exact fractions of the wavelength—one-half, one-quarter, one-eighth—which are nearly as efficient and much more practical. In applications using higher frequencies, transmitting antenna size is not a problem because of the much shorter wavelengths of the signals. UHF TV signals have a wavelength of only a foot or two; the signals received by backyard satellite dishes are only two or three inches in wavelength.

Antennas are constructed in ways that vary transmission patterns and enhance reception. Transmission can be the same in all directions, called omnidirectional, or focused into a single beam for point-to-point transmission. Between these extremes transmission patterns can be varied infinitely to suit the coverage needs of the system in use. Likewise, receiving antennas can be equipped with various kinds of collectors or reflectors to gather and concentrate the signal being received. The parabolic reflectors on satellite receiving dishes are a familiar example.

The Electromagnetic Spectrum

The Usable Spectrum. In the broadest definition the elecromagnetic spectrum ranges from waves with a frequency as low as one cycle per second (1 Hertz) through higher and higher frequencies on toward infinity. However, only a small chunk of this total possible spectrum is routinely used for electronic communication. This smaller portion of the total spectrum is called the radio frequency (RF) spectrum.

The radio frequency spectrum begins at 30 kHz and includes frequencies up to about 300 gHz, a range representing the traditional limits of conventional over-the-air and wire transmission technology. Frequencies above this range take the form of visible light, then radiation; those below it include those used by audio equipment and electrical power systems. Laser-optical systems operate outside the RF spectrum, using visible light. However, this requires the use of optical fiber cables for transmission of the light signals. Thus, if you wish to transmit information over the air or through a conventional wire, modes used by the vast majority of today's electronic communication media, you are still limited technologically and economically to the RF spectrum.

Frequency Bands. Over the years a system has evolved that divides the RF spectrum into seven chunks or bands (see Table 2–1), ranging from low frequency (LF) to extremely high frequency (EHF). The 30 mHz point on the spectrum often is used as an arbitrary dividing point between the higher and lower bands. The term *microwaves* generally refers to frequencies at or above 1 gigahertz and refers to their extremely short wavelengths.

As Table 2–1 shows, familiar communication systems make use of different bands. For example, AM radio broadcasting uses frequencies lying at the lower portion of the Medium Frequency (MF) band, while FM radio and television broadcasting take up space in the Very High Frequency (VHF) and Ultra High Frequency (UHF) bands. Satellite communications use even higher frequencies, microwaves the Super High Frequency (SHF) band and above.

Table 2–1. Bands of the Radio Frequency (RF) Spectrum

Frequency Range	*Band Name*	*Typical Telecommunication Services*
300 gHz		
	EHF Extremely High Frequency	Research, experimental, and military satellite uses
30 gHz		
	SHF Super High Frequency	Commercial satellites, radar, microwave, air navigation
3 gHz		
	UHF Ultra High Frequency	Public safety radio, UHF TV
300 mHz		
	VHF Very High Frequency	VHF TV, air navigation, public safety radio, FM radio
30 mHz		
	HF High Frequency	Shortwave radio, CB, Ham radio
3 mHz		
	MF Medium Frequency	AM radio, maritime and air communication, Ham radio
300 kHz		
	LF Low Frequency	Air and maritime navigation
30 kHz		

Band Characteristics. These and other systems are located in different bands because the frequencies in these bands differ in physical characteristics. Although frequency characteristics do not change abruptly at any point along the spectrum, bands at different locations have quite distinct characteristics. These differences make them more technically suitable for some purposes, less for others. One major difference between bands is propagation, the way the frequencies travel when radiated from a transmitter.

In general, lower band frequencies (below 30 mHz), with their relatively longer wavelengths, can pass through solid objects, tend to follow the curvature of the earth, and require longer transmitting and receiving antennas than higher frequencies. Higher band frequencies (above

30 mHz), with shorter wavelengths, can use shorter, more compact antennas but tend to reflect off solid surfaces and travel only in straight lines—that is, line-of-sight.

Higher frequencies are less susceptible to electrical interference, but they are more vulnerable to problems from solid obstacles and moisture in the air. Anyone who has tried to listen to an AM radio broadcast during an electrical storm is familiar with problems electrical interference can cause lower-frequency transmission systems. Likewise, if you have ever tried to watch a UHF TV broadcast wobbling on the screen as the trees bend back and forth on a windy day, you can testify to the problems physical obstacles can cause higher-frequency transmissions.

Consequently, the frequencies used by a communication system have a profound effect on the system's capabilities and design. Furthermore, the purpose of a system often rules out the use of some bands all together. For example, a radio communication system providing aircraft–to–flight-controller communications cannot be designed to use a band of frequencies highly subject to weather interference (as are some high-frequency bands) or one that requires extremely large antennas (as some lower-frequency bands do). Satellite receiving dishes provide another example. Because satellites use extremely high frequencies (4 gHz and higher), a dish must be installed with an unobstructed, direct view of the satellite's position in the sky. Walls, hills, or even dense foliage will block the transmission.

Spectrum Management. Many different communication services use the spectrum. Their technological capabilities and economic viability often depend directly on the bands they inhabit and the particular characteristics of those groups of frequencies. The number of users of any one band is also limited by the amount of frequency "space" each user needs to operate. Thus, the spectrum can be thought of as highly valuable electromagnetic real estate, much in demand.

Management of this "invisible resource"[1] is a complex and often controversial responsibility. It falls largely to the Federal Communications Commission in the United States and similar central agencies in other countries. How these agencies operate is discussed in Chapter 3. By controlling access to the spectrum and setting the rules for its use, such regulatory agencies determine not only the quality and types of service available in their respective countries but the kinds of new electronic media that will proliferate worldwide.

Notes

1. For a comprehensive overview of the technical and political aspects of spectrum regulation, see Harvey J. Levin, *The Invisible Resource: Use and Regulation of the Radio Spectrum* (Baltimore: Johns Hopkins Press, 1971).

3

Control of Electronic Communication

Although they may differ in the extent of regulation, all nations have established some degree of centralized control over critical aspects of electronic communication. This chapter reviews the fundamental reasons for this control and then briefly discusses some of the most common forms or patterns of regulation before offering a detailed overview of the regulatory system in the United States. The conclusion addresses the special problems regulators face in fostering new electronic media.

Reasons for Regulation

One of the world's first truly international organizations was established out of the need to coordinate electronic communication. The forerunner of today's International Telecommunications Union (ITU) first met in 1865 in Paris. These twenty countries agreed on a uniform system of codes and equipment for international use of the first "new technology," the telegraph. Twenty years later the development of the telephone necessitated a conference in Berlin establishing international telephone standards and services. In another twenty years wireless telegraphy led to ITU agreements institutionalizing that service among member nations. This pattern of technological advance followed by ITU response continued unbroken through the development of broadcasting in the 1920s to the present-day growth of international satellite communication systems.

Not only does the history of ITU agreements and conferences reflect the increasing sophistication of communication technology over the last 120 years, but the very existence of the ITU testifies to the necessity for and perhaps the inevitability of regulation at the international level. Furthermore, the delegations that represent countries at ITU conferences include persons who work in national regulatory agencies, the second of the two major domains of electronic communication regulation: national and international. At both the national and international levels, nations regulate electronic communications out of technical, political, and economic necessity.

Technical Need for Regulation. As far back as the invention of the telegraph, there have always been different technical solutions to any given electronic communication problem. The task at hand and the fundamental scientific principles may be the same, but technical approaches often vary enough to make systems incompatible. For example, the 1865 ITU meeting in Paris was called because the use of different types of telegraph equipment and encoding schemes by various countries was inhibiting the international use and development of telegraphy. The meeting produced an agreement elevating one of these telegraph systems to the status of international standard and relegating the others to instant obsolescence.

Today, at both the international and national levels it remains a truism of electronic communications that the sender and receiver of a message must have compatible equipment. The increased sophistication and complexity of telecommunication systems have made the problem of compatibility even more complex. The fundamental difficulty is that new telecommunication systems tend to appear first at the national level, where either the government or market competition has selected a standard. By the time the question of an international standard arises, incompatible systems are already in place. Without regulation there can be no technical standards; without standards, no efficient system of international communication.

A second major dimension of the technological need for regulation is the physical nature of some communication systems. Radio waves travel at light speed to any device capable of receiving them. They are no respectors of political boundaries or national sovereignty. Consequently each transmitting system must use frequencies different from those used by neighboring systems. No two stations within receiving range of each other can transmit on the same frequency without cancelling out each other due to interference. Such frequency allocation problems—which station uses which frequency, when, and with how much power—forced early broadcasters in the United States to request federal regulation of their industry. It also forces international agreement on a system of spectrum allocation designed not only to prevent interference problems but also to provide for some sort of fair and rational distribution of frequencies among nations and among types of systems or services.

Political and Economic Reasons to Regulate. The most fundamental reasons to regulate electronic communication are technological. However, because of its profound impact on society, states are compelled to regulate for political and economic reasons as well.

It can be argued that the flow of information within or across a nation's boundaries is as fundamentally important as the flow of commodities, capital, or even people. Every state ultimately develops some sort of regulatory policy regarding each of these flows, ranging from strict control to laissez faire. The nature and extent of this regulation reflect a state's fundamental philosophy about the purpose of government and its relationship to citizens.

Thus, in addition to technological objectives, every regulatory system is bound by a political and economic philosophy as well. Moreover, in the day-to-day development of regulatory policy, political and economic objectives may take precedence over technological efficiency, resulting in regulatory decisions that make more sense from a political or economic viewpoint than a technological one. A classic example in the United States was the FCC's decision in 1953 to create a dual set of bands for commercial television broadcasting. When the Commission opened the UHF spectrum to television broadcasting, it left the 107 existing VHF television stations in place rather than cause economic damage to the nascent TV industry. In doing so, however, the FCC put the UHF stations at a technical disadvantage that took two decades to overcome. More recently the FCC declined to select a technical standard for AM stereo, videotext, or teletext, leaving the establishment of a standard to marketplace forces, an action based on regulatory philosophy rather than technological efficiency.

Patterns of Regulation

The worldwide spectrum of political and economic systems has given rise to a variety of regulatory systems. Although no two countries regulate electronic communication services in exactly the same manner, four broad patterns of organization exist.

1. Perhaps the most common approach is direct state control. Telecommunication services are operated directly by government departments. Although a certain department—for example, a state television network—may operate with a degree of administrative freedom, it has no real independence from the central state authorities. Financially, its operating funds come from state appropriation, and whatever service fees it may charge are collected centrally as state revenue.

2. In the second major pattern of regulation, communication services are operated by agencies not immediately under state authority. Although these quasi-governmental agencies are set up by government action, they are given their own legal identity through a charter or other document that may restrict direct government intervention in their operation to some extent. Instead, the agency must answer to a governing board or group of trustees established by the charter. The agency may also enjoy financial independence, supporting itself through user charges of various types that return directly to the agency as revenue.

3. In the third regulatory arrangement, communication service is provided by an agency that more nearly resembles a private company in which the state is a powerful partner. In this case, the private agency or corporation is not legally dependent on the state for its origination or existence, as with the state-chartered agency above. However, under the law, the state may own a significant amount of the agency's stock and be a voting member of its board of directors. It may also have other controlling powers exceeding its actual voting share of ownership.

4. Lastly, the state can turn over the operation of electronic communication services to the private sector entirely and act merely as a monitor or watchdog. Depending on the type of telecommunication service, the state may license as many providers as there are available frequencies, or it may select a limited number of providers to operate as public utilities, offering services under regulated rates. In either case the philosophy is to reduce state control to the lowest level consistent with equitably distributed, high-quality service. Economic incentives rather than state requirements are the primary "engine" of the system. The dominant model for decades has been the U.S. model of this approach.

Within this broad framework are many arrangements for the organization and regulation of electronic communication services, with each country developing its own particular version of these models according to its political and economic situation and philosophy. In general it is fair to say that traditionally, the most common arrangement worldwide has been some form of nationalization. In most countries services are either operated directly by the state or by agencies that are state approved or controlled.

Telephone and telegraph systems in most countries outside the United States are administered by cabinet-level offices referred to generically as PTTs (post, telephone, and telegraph). Two major exceptions are Great Britain and Japan, which privatized their state telephone monopolies and introduced competition in 1981 and 1985, respectively.

Radio and television broadcasting also has experienced some movement away from state monopoly in recent years. Although commercial broadcasting has existed alongside the state system in Japan, Great Britain, and Canada for many years, other countries have only more recently hybridized their broadcasting systems. Examples include Italy, where 2,000 private radio stations and 300 private television stations now coexist with the state-operated three-channel radio and TV service (RAI); and France, which recently added privately owned television channels to its government-controlled system. A trend toward deregulation or commercialization of telecommunication services is evident in these and other industrialized countries, but the great majority of nations still follow the state monopoly pattern.[1]

Regulation in the United States

The regulation of electronic communication in the United States is the result of the complex interplay of private industry, several independent regulatory agencies, and government at the federal, state, and local levels.

The Federal Communications Commission. The nexus of most of this activity is the FCC, established by Congress with the Communications Act of 1934 and granted authority over all forms of radio communication. The FCC is an independent regulatory agency, empowered by Congress to set standards and to make and enforce rules needed to regulate electronic communication within the limits prescribed in the Communications Act.

The FCC accomplishes its work through four major bureaus: mass media, common carrier, safety and special radio services, and field operations (see Figure 3–1). The mass media bureau is responsible for standard and low-power television broadcasting, AM and FM radio, and cable system regulation. Telephone and telegraph regulation fall to the common carrier bureau, as do nonwireline common carrier services, such as cellular mobile radio and radio paging. Marine and aviation radio and mobile radio used by safety and public service agencies come under the

Figure 3–1. Major Bureaus of the FCC

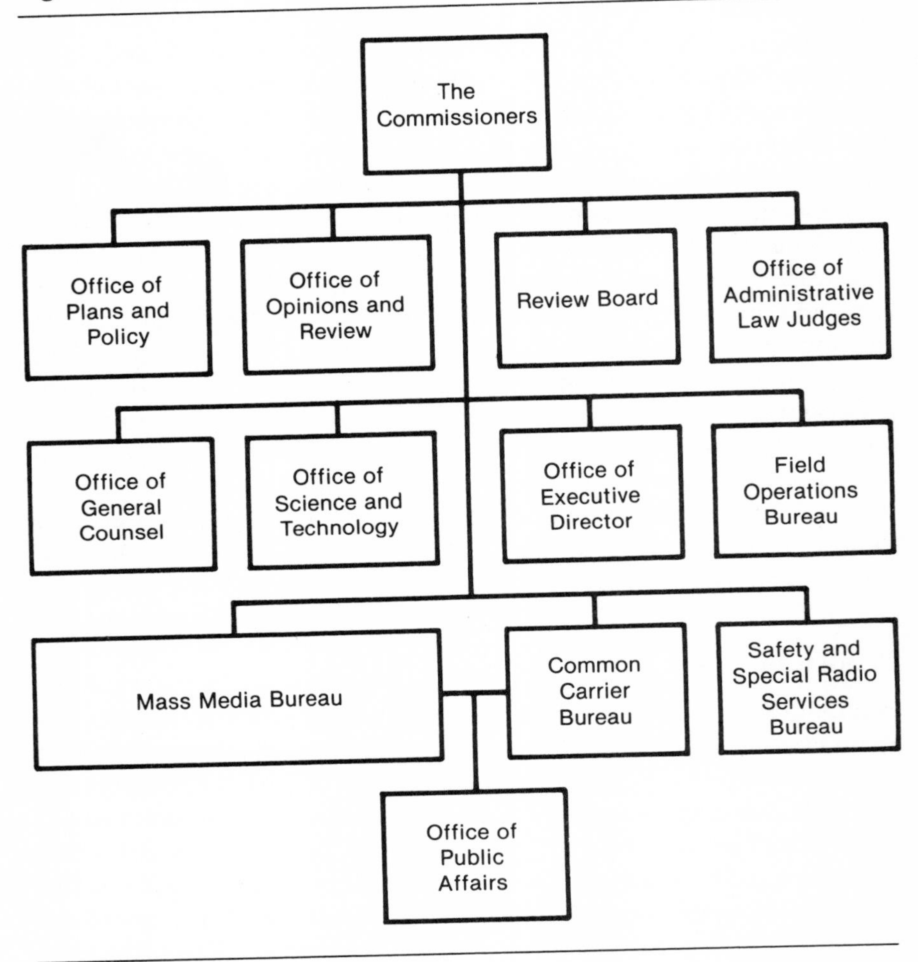

safety and special radio services bureau. The field operations bureau enforces technical standards by inspecting and monitoring radio and television stations and investigates rule infractions.

Various congressional committees oversee the FCC's execution of its mandate by authorizing its funding, approving the president's appointment of its five commissioners, and using their legislative and oversight powers to alternately rein in or prod the FCC. The federal courts referee the regulatory process and have the power to vacate FCC rulings or policies that exceed its authority, are inconsistently applied, or lack a logical rationale.

Other Federal Regulators. Responsibility for communication industry regulation does not rest solely with the FCC. Other federal independent regulatory agencies also deal with electronic media as part of their mandate. For example, the Federal Trade Commission, with the primary task of ensuring fairness in business competition, oversees the use of advertising, including deceptive commercials on broadcasting stations. The Securities Exchange Commission regulates the trading of stocks in publicly held corporations and consequently supervises ownership changes in large communication corporations.

Several agencies in the executive branch play a regulatory or policy-making role as well. For example, the Justice Department's responsibility for enforcing antitrust laws often involves it with the communication industry in significant ways. A Justice antitrust suit led to the breakup of AT&T and the subsequent restructuring of the entire U.S. telephone system, with worldwide implications. The Justice Department over the years has also prevented the sale of a television network, stopped movie companies from forming their own version of HBO, and examined the scrambling of satellite programming services.

In the Department of Commerce the National Telecommunications and Information Administration (NTIA) is responsible for monitoring the government use of the spectrum, evaluating present policies and recommending new ones, and studying technological developments in electronic communication and their policy implications. A final example is the State Department, which represents the United States in international communication matters. Representing U.S. interests on issues like global spectrum allocation and technical standards, the State Department negotiates treaties and other agreements that are often critical to the U.S. communications industry.

Nonfederal Regulation. Although much of the regulation of U.S. electronic communication is carried out at the federal level, state and local government also have roles. In-state and local telephone regulation is carried out by state utility commissions, for example, and although cable systems must comply with federal regulations, their license to operate (referred to as a franchise) is granted by local government. All these government bodies, along with various industry associations and public interest groups, interact to produce communication policy in the United States.

New Technologies: A Regulatory Balancing Act

To a great extent, the availability of new electronic media, regardless of the regulatory system, ultimately depends as much on regulators as on inventors and manufacturers. Fostering the growth of new technologies and services is a complex task that involves important economic and political tradeoffs. While tight regulation and market control may stifle innovation, the opposite extreme, lack of a regulatory framework, can discourage enterprise. For example, lack of technical standards can cause uncertainty in the market and inhibit the development of new services. In the United States a hands-off approach to regulation of new media occasionally has been viewed as counterproductive by developers of the new media. The development of videotext and teletext in the United States has been hampered in part by the FCC's refusal to select a technical standard.

For any regulatory system the challenge is to foster new, more efficient communication services while at the same time preserving the desirable aspects of the existing system. This task is made even more difficult by political and ideological aspects of regulation, which can override technical considerations. In the chapters that follow DBS, HDTV, Cable, STV, and LPTV provide examples of new media whose regulation and development in the United States has been shaped significantly by political as well as technological and economical factors.

Notes

1. Excellent overviews of the political/economic status of European electronic media in the 1980s can be found in Hans J. Kleinstuber, Denis McGuail, and Karen Siune, eds., *Electronic Media and Politics in Western Europe* (New York: Campus Verlag, 1986); Everett M. Rogers and Francis Balle, eds., *The Media Revolution in America and in Western Europe* (Norwood, N.J.: Ablex, 1985); and W.J. Howell, Jr., *World Broadcasting in the Age of the Satellite* (Norwood, N.J.: Ablex, 1986).

PART II
Satellite Communication

4

An Introduction to Communication Satellites

More than 20,000 miles out in space a gossamer string of pearls encircles the earth. A closer look at the dozens of pearls reveals each is an awkward-looking clump of hardware: shiny metal and foil, bristling antennas, and gangly solar panels; some spin, some sit motionless, all apparently are silent. But through each satellite flows a torrent of information: thousands of telephone conversations, streams of computer data, and television signals—all vital to the governments, industries, and mass media on the planet far below. In less than thirty years communication satellites have become an indispensable global electronic medium, touching the lives of people the world over. This chapter reviews their evolution and provides an overview of their operation.

Early Communication Satellites

In October 1957 the USSR launched Sputnik ("fellow traveler"). Only twenty-two inches in diameter, Sputnik was the first artificial satellite to orbit the earth. Although not a communication satellite, Sputnik did transmit a continuous "beep-beep" signal from a battery-powered radio as it circled the globe in a short-lived, low orbit. Sputnik was followed the next year by Explorer, the first U.S. satellite, which was used for scientific research. Sputnik and Explorer not only heralded the U.S.–USSR space race with its enormous military and scientific implications, but also were the forerunners of today's communication satellites.

During the early 1960s as larger and more powerful rockets were developed to launch them, the size and capabilities of all types of satellites increased rapidly. The first communication satellites were purely experimental. For example, in 1960 the United States orbited Echo I, a 100-foot-diameter balloon, to investigate the possibility of literally bouncing radio messages from one earth station to another. Other satellites were launched that electronically relayed earth signals, but they were in orbit for only days or weeks.

In 1962 AT&T's Telstar provided the first live telephone and television relay service between Europe and the United States. Unlike its experimental predecessors, Telstar remained in orbit a year and provided a commercial service to the public. Telstar peaked public interest in communication satellites and even inspired a popular record of the same name. It was followed within months by a similar system, RCA's Relay. Although they each weighed only about 200 pounds, Telstar and Relay resembled today's communication satellites because they used solar panels and rechargeable batteries for power and actively amplified and retransmitted the signals they received.

They had two very important limitations compared to their modern successors, however. Their small size and low power severely limited the number of telephone conversations or television signals they could carry. More important, their low, fast-moving orbit meant they were visible from earth stations for only a few minutes every couple of hours. Transatlantic telephone conversations or television transmission could take place only while Telstar or Relay were passing over the Atlantic. During this time, the earth station antennas had to follow, or track, the moving satellites to maintain the electronic link, something like playing handball off a moving wall—with your shoes stuck to one spot.

The first satellites to solve this problem were Syncom I and II, launched by the United States in 1963. The Syncom "birds" were the world's first geostationary satellites, prototypes of all the communication satellites in orbit today. As explained later in this chapter, geostationary satellites orbit much further out than low-orbit satellites and appear motionless from the earth. Less expensive, immobile earth station antennas could be used with the Syncom system because there was no need to track the satellites. Transmission was possible around the clock instead of a few minutes every couple of hours. In 1964 a third Syncom over the Pacific relayed dramatic live television coverage of the Tokyo Olympics back to the United States. The geostationary Syncoms had initiated the age of practical intercontinental satellite communications.

Today's Global Satellite Systems

While early U.S. efforts were demonstrating the advantages of geostationary systems for international telecommunication, groundwork was being laid for a global satellite network. In 1964 the United States and eighteen other United Nations members established INTELSAT, the International Telecommunications Satellite Organization. The first INTELSAT satellite, Early Bird, was placed in geostationary orbit over the Atlantic in 1965. More INTELSAT satellites soon followed over the Atlantic, Pacific, and Indian oceans. By summer 1969 the INTELSAT network was developed to the point that half a billion viewers worldwide were able to watch live as Neil Armstrong set foot on the moon.

In the 1970s INTELSAT continued to grow, adding members to the cooperative and investing billions of dollars in satellites and earth station technology. Today INTELSAT is the world's dominant international satellite transmission service, with sixteen powerful satellites linking over a hundred member countries. It handles about two-thirds of the world's telephone traffic and a significant portion of the video transmission. In most cases INTELSAT sells its services at wholesale prices to the government-run telephone systems of its member countries, which own shares of the organization. In the United States, where telephone services are provided by private companies like AT&T, INTELSAT membership is through COMSAT, a government-chartered corporation that has been one of the largest shareholders in INTELSAT since its creation.

Two other much smaller global satellite systems exist. INTELSAT was joined in 1971 by INTERSPUTNIK, a similar cooperative established by the Soviet Union to serve Eastern European and other nonaligned countries around the world. In the late 1980s INTERSPUTNIK had about fifteen members linked by several Gorizont satellites over the Eastern Hemisphere.

The third global satellite system, also a cooperative, is INMARSAT, established in 1979. With satellites over the Atlantic, Pacific, and Indian oceans, INMARSAT provides maritime communications for the shipping fleets, ocean liners, and offshore oil rigs of about forty member countries, including the United States and the Soviet Union.

Domestic Satellite Systems

While global satellite cooperatives were being developed in the 1970s, geostationary satellite systems used exclusively for national communications also began to appear. In 1965, with the launch of its first Molnya satellite, the Soviet Union established the world's first domestic satellite communication system to relay data, voice, and video transmissions across the USSR. However, neither the original Molnya satellite nor its present-day successors are geostationary, because much of the USSR is too far north to be served by satellites in fixed position over the equator.

The world's first geostationary domestic satellite system (domsat) was developed by Canada. In 1972 Canada launched Anik ("little brother"), the first in a series of geostationary satellites used for telecommunication across the vast Canadian provinces. That same year the U.S. Federal Communications Commission announced its "open skies" policy, permitting operation of geostationary satellites by private corporations.

Westar, the first U.S. domestic bird, was orbited by Western Union in 1974, then came RCA's Satcom I in 1975 and Comstar in 1976. Each was the first in a series of satellites by the same name. The Comstar series was used initially for telephone transmission by AT&T; Satcom I and its successors have been best known as television transmission birds for the U.S. cable and broadcasting industries.

Throughout the 1970s the Anik, Westar, Satcom, and Comstar satellites were joined by increasing numbers of U.S. domsats in the skies over the Western Hemisphere. Today many other nations worldwide have launched similar national systems. Brazil and Mexico have joined the United States and Canada in the West, and most major Western European nations, the USSR, China, Japan, India, Indonesia, Israel, and Nigeria all have national domsat systems. More than 100 domsats were in orbit worldwide in 1988, and many more were planned by these and other countries.

Regional Systems

In addition to the national systems, a number of regional systems arose in the 1980s, developed by neighboring countries on a cooperative basis. For example, EUTELSAT, founded in 1977, consists of about twenty Western European countries, including the United Kingdom, West Germany, France, and Italy, which together own over 50 percent of the consortium's shares. Three geostationary EUTELSAT birds offer telephone, video, and data links to the Western Europe/Mediterranean area. Two other important regional systems are ARABSAT, set up by seventeen Middle Eastern nations, and PALAPA, a system serving the thousands of islands comprising Indonesia as well as neighboring Southeast Asian countries. Although relative latecomers to the international communications arena, such regional systems represent the first challenge by its own members to INTELSAT's dominance of the global satellite communication market.

In summary, the entry of regional systems has added a third component to what has become a trilevel system of global satellite communications services. The first step was the organization of INTELSAT in the late 1960s, leading to the establishment of a truly global international satellite communication system. During the 1970s and 1980s this global system was joined quickly by national systems in most industrialized countries. More recently, regional systems have provided an intermediate level of service. From a single Syncom satellite over the Atlantic in 1963 to about 150 geostationary satellites encircling the globe, the growth of communication satellite technology in the last twenty-five years has been a remarkable human achievement.

Orbits: Geostationary and Other

Many satellites orbit the earth in an egg-shaped path known as an elliptical orbit. At the point of the orbit farthest away from earth (the apogee) the satellite appears to be moving most slowly as seen from the earth. At the short end of this elliptical path, when the satellite reaches the point closest to the earth's surface (the perigee) it appears to be moving the fastest.

Satellites in an elliptical orbit appear, from the earth, to rise at one point on the horizon, cross the sky in an arc, and disappear at another point on the horizon. Depending on the location of the orbit, the satellite's path across the sky runs from north to south, east to west, or any points in between. Satellites in low orbit near the earth pass overhead quickly, in as little as ten or fifteen minutes. Those in higher elliptical orbits typically are in view for several hours.

Strictly speaking, any orbit that takes twenty-four hours to complete is referred to as a geosynchronous orbit, regardless of its shape or angle. The orbit is geosynchronous because the satellite completes one trip around the earth in the same amount of time it takes the earth to complete one turn on its axis. In 1945 British writer Arthur C. Clarke, later the author of *2001: A Space Odyssey* and numerous other books, suggested in a paper that a special type of geosynchronous orbit would allow just three satellites to provide twenty-four-hour-a-day microwave relay service for the entire planet.[1]

Clarke calculated that a satellite in a circular—not elliptical—geosynchronous orbit 22,300 miles directly above the equator would, to an observer on the earth, appear to hang motionless in the sky. From its distant viewpoint this geostationary satellite could "see" a bit more than a third of the earth below it. Three could remain in sight of each other constantly and relay radio signals from nearly any point on the planet to any other—a nearly instantaneous global communication system. This unique circle around the earth, the "parking" place for the world's geostationary satellites, has become known as the GSO (geostationary satellite orbit) or the Clarke Belt.

Satellite Launch and Positioning

After Clarke pointed out the possible advantages of geostationary orbit, it took over twenty years to overcome the technological difficulties involved in getting the necessary hardware from here to there. Although a common occurrence today, the launch of a geostationary satellite is still a complex, expensive operation and not without an element of risk. The launch vehicle itself may fail, as have NASA's space shuttle and Delta rocket, the U.S. Air Force's Titan rocket, and the European Space Agency's Ariane rocket, or the satellite can malfunction. For example, RCA's Satcom III disappeared shortly after launch in 1979, the first casualty among commercial domsats. By 1988 the loss of several others had driven up the cost of insurance to the point that some satellites were being launched with partial coverage or none at all.

As Figure 4–1 indicates, the trip to the GSO entails several basic steps. After your satellite has been constructed, it is delivered to the launch contractor of your choice. In 1988 possibilities included not only NASA in the United States or Europe's Arianespace but also newly developed commercial launch services in Japan, China, and the USSR. With NASA you may choose to use either the space shuttle or an unmanned rocket, referred to as an expendable launch vehicle (ELV). All other launch services use ELVs exclusively. With its sixty- by fifteen-foot cargo hold, the shuttle can launch unusually large satellites or multiple satellites simultaneously. ELVs are limited to carrying conventional-size birds, one at a time. Regardless of the vehicle used, the steps are essentially the same.

Figure 4–1. Steps in Launching a Communications Satellites

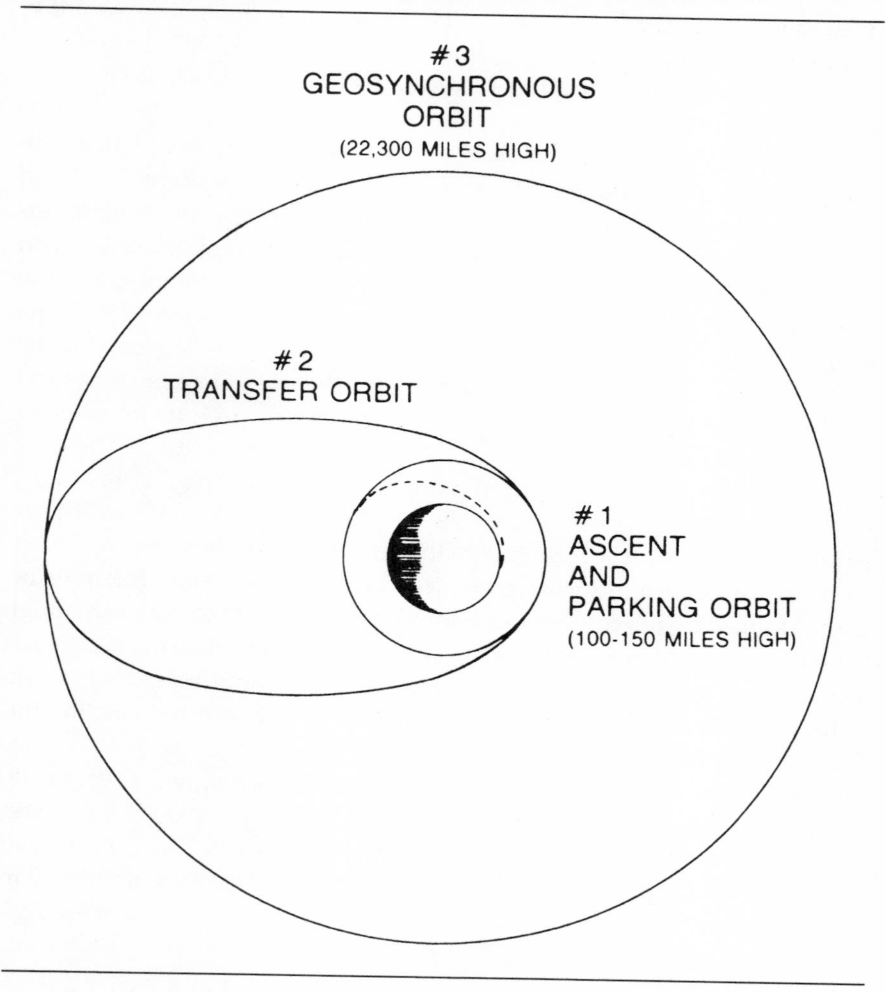

Most of the energy expended in a launch is used up in the first step, the ascent, a ride from the launch pad to a low orbit. The satellite must be be constructed to withstand the enormous vibration and G-forces of the ascent as the vehicle accelerates from a standstill to 17,000 miles per hour in less than ten minutes. With the shuttle, the ascent ends with entry into a low orbit about 150 miles high. From this low parking orbit, the shuttle crew simply opens the cargo bay and dumps your satellite overboard, giving it a spin for stabilization. If launched by an ELV into a parking orbit, the satellite sheds its protective housing and frees itself from the launch vehicle on command from the ground. On its own, moving over 17,000 miles per hour relative to the earth below, the satellite emits a radio signal that allows it to be located and tracked from the ground.

Next, the satellite must be transferred from parking orbit to its allocated spot over the equator 22,000 miles farther out. To do this, ground controllers fire a rocket engine on the satellite for a few minutes, boosting it into an elliptical transfer orbit 22,300 miles high. This boost is called the *perigee kick* because the kick from the rocket motor comes while the satellite is still close to the earth, at the perigee of the transfer orbit. Some ELVs, like the Ariane, bypass the parking orbit and fly the satellite directly to the correct transfer orbit altitude without the need for a perigee kick.

Boosted to the correct altitude, the satellite must be given another kick to change the shape of the orbit from elliptical to circular. The perigee kick motor or another is fired momentarily, providing what is called the *apogee kick*. At the same time, the satellite must be lined up with the equator and flown to its assigned slot on the Clarke Belt.

Transfer orbit is the step in which the satellite is most likely to be lost. It is easy to imagine how any miscalculation or misfiring during the perigee or apogee kick could spin the satellite into a useless orbit or cause it to be lost completely. Potential problems are compounded by the fact that only a limited amount of fuel is available on board the satellite for these maneuvers.

If all goes well with the apogee kick and other adjustments, the satellite finally achieves geostationary orbit at its assigned parking slot on the Clarke Belt. Although now apparently motionless to observers below, the satellite is traveling through space at about 6,900 MPH. Once its solar panels and antennas are deployed and its subsystems turned on and checked out, the satellite begins operations.

Station-Keeping and Drift

Station-keeping refers to the task of keeping a geostationary satellite properly operating in its correct position with its antennas aligned correctly with the earth stations below. A number of systems on the satellite work together to make this possible. The power system includes solar panels, which provide primary power, and rechargeable batteries for when the sun is not available. The altitude control system consists of sensors that tell the satellite where it is in relation to the earth, stabilizing devices like spinning balancing wheels, and small gas jets to make needed corrections. A temperature control system must cope with the satellite's tendency to heat up on the side facing the sun and freeze on the opposite side. The main communication system includes antennas for receiving and transmitting signals from earth, and transponders, which amplify the signals and shift them from their inbound to outbound frequencies. Another communication system (TTC) provides earth-bound controllers with information to monitor the satellite's position and functioning (tracking and telemetry) and the means to control both remotely.

All these systems come into play when coping with the satellite's tendency to drift from its assigned position. Drift is caused by the gravitational effects of the sun, moon, and earth on the satellite, which tend to pull it out of its orbit. Also, because the earth is not perfectly round, its pull on the Clarke Belt satellites is not uniform. Each hemisphere has a point of highest and lowest pull, with gradations in between. Satellites tend to drift or "slide" toward the point at which the pull of gravity is the strongest in their hemisphere, located roughly over India in the East and over the United States in the West. The pressure of solar radiation also contributes to drift, although much less so than gravitational effects.

Communication Via Satellite

Channels and Circuits. Communication satellites are the space version of the familiar microwave transmission tower on earth. They merely retransmit microwave signals aimed at them from transmitting antennas below to receiving antennas in different locations. Generally, transmissions to and from satellites are called uplinks and downlinks and come in a variety of combinations. A one-way link from one earth station via satellite to a receiving earth station is called a channel. For example, the link between a subscription movie service, like HBO, and one of its cable system affiliates is a channel. A full two-way link between two earth stations is known as a circuit and requires two channels, one in each direction. A telephone conversation requires a circuit.

The Transmission Industry. Terrestrial facilities for uplinking and downlinking satellite signals are owned or leased by users. In the 1980s a wide range of businesses evolved to provide satellite-related equipment and transmission services, from the local backyard dish dealer to companies that purchase transmission equipment, lease transponder time, and then sell communication service to retail users. Among these are an increasing number of teleports.

A teleport is a satellite up- or downlinking facility built specifically to provide a telecommunications connection between local businesses and communication satellites. It consists of a collection of transmission and reception antennas erected in a specially shielded area, along with the electronics needed to connect local users by terrestrial lines to the facility. Most users are businesses in metropolitan areas needing high-capacity voice and data transmission services. According to a 1987 estimate, sixty-eight major international teleports were operating worldwide, with more than 200 expected by 1999.[2]

Frequencies. Satellite transmission takes place over frequencies set aside by the telecommunication authorities in each country. These agencies use the frequencies that have been allocated to their countries by international agreements worked out through the International Telecommunications Union.

The most widely used set of frequencies is called the C-band. C-band frequency assignments cluster around 4 gHz for downlinks and 6 gHz for uplinks. Uplinks and downlinks use separate frequencies to avoid interference between inward- and outward-bound transmissions. For example, home satellite dishes in the United States receive programs on downlink frequencies between 3.7 and 4.2 gHz. The programs are transmitted by subscription services on uplink frequencies between 5.9 and 6.4 gHz. All these up and down frequencies are in the C-band. C-band transmissions are very weak and require an eight- to twelve-foot dish for adequate reception. The second most widely used band is the Ku-band, a much higher set of frequencies clustering around 12 gHz. Ku-band transmissions can be received with a smaller dish (three to six feet) but are more subject to weather interference.

Satellite Capacity. Satellites relay transmissions, shifting them between assigned uplink and downlink frequencies with their transponders. The number of transponders on a satellite and their capacities determine the satellite's total communication capacity. Most commercial satellites in operation today have no more than twenty-four transponders, with some older ones limited to twelve. Satellites with as many as forty transponders are expected to be commonplace in the near future. As a rule of thumb, a single C-band transponder can relay one color television signal, or a dozen commercial radio station signals, or 1,200 telephone conversations. In 1988 orbiting U.S. domsats had an estimated total capacity of more than 500 transponders.[3]

Service Categories. The great variety of satellite communications fall into four broad regulatory categories. Broadcast Satellite Service (BSS), the relay of radio and television signals to broadcast stations and cable systems; Direct Broadcast Service (DBS), high-powered satellite transmission of programs direct to subscriber homes; Fixed Satellite Service (FSS), relay of telephone, data, and other signals to a limited number of designated receivers, usually for further terrestrial distribution; and Mobile Satellite Service (MSS), communication links for ships at sea, aircraft, and terrestrial vehicles. The great majority of global satellite communication is in the FSS category, followed by BSS. DBS (see Chapter 5) is still in developmental stages and MSS consists primarily of maritime services.

Advantages of Satellite Transmission

The primary advantages of geostationary satellites as transmission facilities are derived from their enormous signal coverage area, or footprint, which can include as much as a third of the planet beneath them. This coverage enables them to span distances at a fraction of the cost of traditional terrestrial transmission technologies. For example, using telephone lines and terrestrial microwave relay towers, it is much more expensive to build a system linking New York to Los Angeles than one linking New York to Boston; the longer the distance, the more equipment must be used.

In contrast a single satellite covering North America can relay calls from New York to Los Angeles or to Boston, using the same equipment, regardless of distance. Thus, distance does not affect the cost of satellite transmission. Furthermore, a satellite transmission can be received simultaneously in multiple locations in the coverage area without affecting transmission cost. Terrestrial transmission to many locations requires the construction of a network of lines and relay towers connecting each location. These two advantages over terrestrial transmission—insensitivity to distance and unlimited locations—are the primary economic incentives to build, launch, and use satellite systems.

Disadvantages and Limitations

Transmission Delays. Inherent in the advantages of space-relayed communications are some drawbacks as well. Although satellite transmissions span great distances economically, they must always travel great distances regardless of how close together the earth stations may be. Even traveling at the speed of light, the nearly 50,000-mile trip made by a one-way satellite transmission takes nearly a third of a second—about a quarter of a second travel time plus a bit more processing delay on the satellite and at the earth stations. Terrestrial systems have no comparable delay.

In a two-way satellite transmission like a telephone call, it takes nearly two-thirds of a second longer to ask a question and receive an answer than it does using terrestrial relays. In conversation this is rarely more than an inconvenience. But not so in the case of high-speed data transmission, like a conversation between two computers, in which millions of bits of data are transmitted in a second. The computers' optimum transmission speed greatly exceeds the delay, which can cause significant problems and place limitations on transmission speed not necessary when using terrestrial communication systems.

Antenna Size. Because of the size and weight limitations on satellites dictated by the economics and capabilities of launch vehicles, they can transmit only very low-power signals. Consequently satellite transmission requires inconveniently large receiving antennas, which have limited use in some circumstances. Great progress has been made in increasing satellite power and reducing earth station size, but the present affordable state-of-the-art has distinct limitations. You can't, for example, mount a typical home satellite dish on the roof of a typical home—yet.

Security. Because satellite signals cover a large area, they can be readily received by anyone in the coverage area, complicating security problems for certain kinds of transmissions. Even when they are scrambled, satellite transmissions are less secure than scrambled terrestrial transmissions because the eavesdropper can operate from anywhere in the vast coverage area. The terrestrial eavesdropper has to intercept the signal somewhere along the line of transmission.

Furthermore the satellite itself is a vulnerable target for jamming. Most commercial communication satellites are "dumb" in the sense that they do not know who is sending up the signals they are relaying—only whether the signals are on the correct frequencies. Using a transmitter more powerful than the authorized sender, a pirate or saboteur can easily jam a legitimate transmission or override it with a bogus message. This potential for interference came to light in the United States in 1986 when a disgruntled satellite dish dealer jammed a Home Box Office transmission, overriding the normal movie programming for a few seconds with a message protesting subscription charges. The episode led to an FCC inquiry into the use of a mandatory identification system for all satellite programming originators and tighter security at commercial transmission facilities.[4]

Frequency Crowding. Because they travel through space and the atmosphere, satellite signals are subject to numerous other types of interference, natural and artificial. Such interference is sometimes noticeable as reception problems on cable TV systems or noise on telephone lines. One contributor to interference is frequency crowding. C-band satellites use the same frequencies as terrestrial microwave transmission systems. Interference from frequency crowding has become a major factor in the placement of earth stations in major metropolitan areas, with many dishes being constructed in pits or surrounded by earthen or concrete walls to shield them from other transmissions.

Weather. The proliferation of C-band systems and subsequent frequency crowding has led to a newer generation of satellites using a higher frequency band, called Ku (12 to 14 gHz), which can use somewhat smaller dishes and is not used for terrestrial transmission. However, Ku-band transmissions are much more susceptible to weather interference than C-band and can be interrupted by heavy rains and thunderstorms. Consequently, in addition to the C- and Ku-band satellites in orbit, some use both, and as the newer Ku-band becomes more crowded, there are experiments with even higher frequencies. Japan has experimented successfully with data transmission using the Ka-band (20 to 30 gHz).

Solar Eclipses. Finally, the sun places certain limitations on the use of geostationary satellites. First, there is the problem of the eclipse of the sun by the earth. Every earth day there is a time when the satellite finds itself on the dark side of the earth because the part of the earth below the satellite has turned away from the sun. During much of the year the angle between the earth, the sun, and the satellite is such that the satellite can still see the sun at night because of its great altitude.

However, during two annual periods the satellite, the earth, and the sun line up in such a way that the satellite cannot see the sun when it passes behind the earth. For a few weeks in March and September all geostationary satellites lose solar power daily for up to seventy minutes during the nighttime below and must rely on their batteries to continue operation. Because no batteries can be recharged indefinitely, this annual blackout is one reason that satellites last only seven to ten years—their batteries wear out.

Sun Outages. Also at this same time of year, when the sun, earth, and satellite are in a straight line, the satellite finds itself positioned once a day with the sun directly behind it and the earth directly in front. The receiving dishes below, aimed directly at the satellite, are also aimed directly at the sun. The sun emits enough electromagnetic radiation to interfere with the reception of the satellite's weak signal, a problem called a sun outage. During sun outages signals are rerouted to other satellites, or alternate receiving dishes are used. Most television viewers have heard announcements during September and March that sun interference is causing momentary transmission problems.

Conclusion: The GSO as an International Resource

The geostationary orbit (GSO) is among the world's unique resources. Because of the laws of gravity and the size and shape of the earth only one orbit is suitable for geostationary satellite communications. This uniqueness is compounded by certain inescapable limitations on its use. Although the full 360 degrees of the GSO or Clarke Belt traverse 165,000 miles of space, only limited portions of it lie over the earth's great continents. For example, only about a seventy-five-degree arc of the GSO is usable for most North American domestic communications. Within each of the several usable arcs worldwide the number of satellites is further limited by the need for spacing to prevent signal interference.

The world's first generation of C-band domsats, launched during the 1970s, were placed in parking slots about four degrees or 1,500 miles apart on the GSO. Increasing demand for parking slots has led to three-degree spacing, which is rapidly giving way to two degrees, made possible by more sophisticated satellites. The GSO is becoming crowded as nations rapidly fill up the portions that can be used for domestic communication. GSO crowding is a major concern, especially among nations that have not yet occupied a piece of this valuable extraterrestrial real estate.

Limitations on GSO space have two potential solutions, technological and political. From a purely technological standpoint several remedies are available. More powerful satellites with more selective reception equipment can be built and are being used. New bands of frequencies can be used, avoiding interference with existing bands. Satellites can share the same parking slot, taking turns using different frequencies. Eventually satellite "farms" can be built: unmanned platforms with enormous solar collectors shared by dozens of satellites that can be individually flown to the farm and then set loose when they are obsolete.

However, technological solutions all come with a price tag; more sophisticated satellites are more expensive. Expense is a critical issue in commercial satellite communications because the current demand is based largely on its economics. For domsat builders and operators, there is a tradeoff between what is technologically possible and what is economically feasible. Technological improvements will be introduced only to the extent that the low cost of satellite transmission can be sustained. The increasing use of fiber optic trunk lines for terrestrial telephone communications has already demonstrated that satellite technology is not invulnerable to competition from other transmission systems. In other words, technological solutions to GSO crowding have economic limitations.

Political solutions to the shrinking GSO involve questions of international equity. If, in fact, the GSO is a limited international resource, then should it be used primarily by those nations that had the resources to get there first? The fundamental question of fairness in allocation of parking spaces is addressed by the International Telecommunications Union in regional and world conferences. As the limitations on the GSO have become more apparent, many less developed nations have expressed concern that they are being left out of the "land rush" and have become dependent on industrialized countries for transmission facilities. Some have proposed an allocation system that would reserve GSO space for a country regardless of whether it is ready to make use of it immediately. The United States among others, supports the present policy in which countries with satellites ready to orbit are given preference in space allocation—something of a first-come, first-served system.

As worldwide demand for satellite transmission facilities continues to expand, the technological sophistication and capabilities of communication satellites will continue to increase. The governments, media, and businesses across the globe will find even more uses for this technology, and our dependency on satellite transmission systems will increase. So too will the value of space in the GSO and the political difficulties involved in developing an equitable international plan for its use. Ultimately, the realization of geostationary satellite technology's full potential for providing global communications will depend more on political and economic considerations than technological.

Notes

1. The original paper submitted to the British Interplanetary Society in 1945 ("The Space Station: Its Radio Applications") is reprinted in *Spaceflight* (March 1968): 85.
2. Lois Bruu, "Teleports Start Fulfilling Promise: Bustin' Out All Over," *MIS Week* (23 April 1987): 22.

3. "Satellite Transponder Supply and Demand," *Multichannel News* (15 June 1987): 20.
4. See Craig Leddy, "Capt. Midnight Surrenders Just as Authorities Close In," *Electronic Media* (28 July 1986): 8.

5

Satellite-to-Home Broadcasting

In 1988 nearly 2 million U.S. homes, hotels, and bars received television programming direct from satellites using bulky backyard receiving dishes, and the number of dish owners was increasing at a steady pace. Satellite-to-home broadcasting is a reality. But this reality is far from the high-tech version of satellite-TV originally envisioned by certain entrepreneurs and governments: tens of millions of homes receiving programs via inexpensive umbrella-sized dishes. Although high-power satellite-to-home broadcasting is on hold in the United States, several European countries have embarked on ambitious satellite broadcasting ventures.

Pioneering Home Satellite Reception

Early geostationary communication satellites were not developed specifically for television broadcasting but were used primarily for intercontinental telephone transmission (see Chapter 4). By the early 1970s the little commercial television programming that was on satellite in the United States was limited to internal transmission (backhaul) of international news or sporting events by TV networks. In Western Europe the situation was much the same. Satellite receiving dishes were very large, ten meters or more, and extremely expensive, starting at $100,000.

However, lack of programming and the cost of technology did not discourage the world's first backyard dish builders. For example, in December

1975 BBC technician Steve Birkill used a homemade antenna and receiver to successfully pick up video from India. Birkill's dish received pictures from a satellite being used by the Indian government to relay television programming to isolated rural villages.[1]

The Western Hemisphere's first backyard dish arrived a year later. In fall 1975 HBO, the cable pay movie service, began transmitting programming to cable systems using RCA's Satcom I satellite, launched a few months earlier. Not only did HBO's start-up eventually revolutionize the cable industry (see Chapter 12), it also inspired a Stanford University electrical engineering professor to build a homemade dish and receiver using salvaged microwave equipment. According to industry lore, in September 1976 Professor H. Taylor Howard tuned in HBO on a homemade rig and became the first person in North America to receive TV programming at home direct from a satellite.

Deregulation and Growth

For the next few years satellite programming services proliferated, but viewing by other than cable subscribers was limited to a handful of experimenters who could design and operate homemade receiving equipment. Although prices were dropping, commercially available dishes were very expensive, in part due to Federal Communication Commission rules. Licenses were required for all satellite receiving antennas, and technical requirements stipulated that they be quite large in order to produce an acceptable picture.

As manufacturers rapidly improved the performance of commercial satellite receiving equipment, FCC technical restrictions were reduced. In 1979 the FCC dropped the licensing requirement for television receive-only (TVRO) dishes. Freed from the licensing requirement and spurred by rapidly dropping equipment prices, the backyard dish industry was born. Thousands of TV viewers living in uncabled, mostly rural areas invested in home satellite receiving equipment.

Early buyers paid $5,000 to $10,000 to view all the cable programming services free. They could also eavesdrop on a variety of other satellite-relayed programming, including TV network sports and news and even private teleconferencing transmissions; in all, more than 100 programming sources were available for the taking. Although the legality of viewing satellite TV services without subscribing was uncertain, retail dish sales soared, and dealers formed an industry association. With equipment continuing to improve technically and prices dropping to the $2,000 level, sales averaged about 50,000 dishes a month nationwide by 1985. Dishes appeared in suburban areas where TV and cable were readily available.

HBO Pulls the Plug

The legal status of backyard dishes was clarified somewhat by a 1984 amendment to the Communications Act of 1934. The law recognized the right of persons to receive satellite broadcasts but also authorized programmers to scramble transmissions and charge for their reception.[2] In January 1986, after distributing decoders to its cable system affiliates, HBO began scrambling. HBO's scrambling system used the VideoCypher II decoder, which took four years and $10 million to develop. Backyard dish owners across the nation could no longer view HBO without buying the $400 decoder. It rapidly became the de facto industry standard as the other pay-TV services, superstations, and advertiser-supported services followed HBO's lead. By 1988 perhaps 100 unscrambled program sources remained, but virtually all the most popular services, including the three commercial broadcast networks, were scrambled.

The Scrambling Controversy

Scrambling hit the satellite dish industry hard in 1986. Thousands of dealers went out of business as sales plummeted to a fraction of the 1985 boom level. For several months a shortage of decoders for sale frustrated suspicious dish owners and dealers. Then, as decoders became available, a slow recovery began. By 1988 dish sales were back to about half the boom level, 25,000 per month, and 300,000 dish owners had installed descramblers.[3]

Despite the partial recovery in dish sales, the scrambling and subscription system became the focus of a controversy involving the home dish industry, dish owners, programmers, and the cable industry. In addition to selling subscriptions directly to dish owners, programmers had authorized their cable affiliates to sell programming to dish owners in and near their franchise areas. Dish dealers and owners accused the programmers of colluding with the cable industry to monopolize program distribution. Dish owners also made allegations that they were being overcharged for programs available at lower cost to cable subscribers. Congressional hearings were held on pricing and access to satellite programming raising the possibility of federal legislation. However, inquiries by the Justice Department and the FCC failed to corroborate the allegations. The FCC concluded in 1987, a year after scrambling began, that prices for scrambled services had dropped somewhat and that no evidence indicated that programmers or cable operators were treating dish owners unfairly. The FCC also pointed out that the recent appearance of program packagers in the industry was a positive development, introducing healthy price competition.[4] Packagers buy several services from different programmers at wholesale rates, then retail them as a package to dish owners, at a lower total cost than separate subscriptions.

DBS versus Backyard Dishes

Backyard dishes are aptly named because their size, typically eight to twelve feet, effectively prevents them from being more conveniently installed on a rooftop or strapped to a chimney like a conventional TV antenna. Their size and need for stability in high winds require they be mounted in a concrete footing. The unwieldy dish is necessary to collect and concentrate the extremely weak signal used for television transmission by most satellites. To reduce dish size and expense a much more powerful and expensive satellite is needed.

As early as 1977 the International Telecommunications Union (ITU) anticipated the development of such specialized, high-power TV satellites and set aside Ku-band frequencies for a service designated DBS (direct broadcast satellite). Broadcasting to homes using conventional C-band satellites and large dishes—the existing system in the United States—became known as low-power satellite broadcasting.

The Comsat DBS Proposal

High-power satellite-to-home broadcasting or DBS was proposed by Comsat to the FCC in 1979. Through a subsidiary, Satellite Television Corporation (STC), Comsat proposed developing a satellite large enough to deliver a Ku-band TV signal 100 times more powerful than conventional satellites. This signal would produce a satisfactory TV picture using an easily installed, umbrella-size rooftop dish antenna costing only a few hundred dollars. Four such satellites, each with four transponders, would cover the country with four channels of programming at a total cost of nearly $1 billion.

In 1982 the FCC announced rules governing DBS service and began accepting license applications.[5] STC's license, the first U.S. DBS license, was granted in September. By early 1984 fourteen more DBS licensees had joined STC, all with FCC-approved preliminary plans for DBS using high-power satellites and small receiving dishes.

The Medium-Power Challenge

Between 1979 and 1984, while Comsat and others were developing plans for DBS, existing TVRO dish technology gradually improved. Conventional satellite power increased somewhat, and newer dishes, although still large, became less bulky. While Comsat and other DBS licensees pressed on with their plans for high-power satellite broadcasting, a company called United Satellite Communications Inc. (USCI) leapfrogged them all, in a sense, by inaugurating a third version of satellite-to-home TV.

Using an existing Ku-band Canadian satellite that was somewhat more powerful than the U.S. C-band birds being watched by the large backyard dishes, USCI was able to offer subscribers five channels of pay programming using dishes only three to five feet in diameter. It was neither true high-power DBS nor low-power C-band satellite TV but something in between, and it could operate without a DBS license. Starting in November 1983 USCI eventually signed up more than 11,000 subscribers in the Northeast and Midwest. But in fifteen months USCI went off the air and declared bankruptcy, unable to offset its expensive start-up costs with adequate income.

USCI's short-lived success demonstrated that DBS-type service could be provided to medium-size dishes without the expensive high-power DBS-only satellites. Facing possible competition from later USCI-type companies and watching many potential customers purchase large dishes rather than wait for DBS, Comsat dropped its DBS plans in late 1984. Seven other licensees also either withdrew or were dropped by the FCC for lack of progress that year. The withdrawal of Comsat, the first and most prestigious licensee, signaled a decline in the near-term prospects of high-power satellite broadcasting in the United States.

Since then there have been a number of active DBS applicants and several licensees with announced plans, but U.S. DBS prospects remain uncertain. Growth of the low-power, C-band version of satellite TV, the demonstrated feasibility of medium-power Ku-band satellite broadcasting to medium-size dishes, and the major problem of programming expense have combined to slow DBS development.

Satellite Receiving Equipment Features and Operation

Regardless of whether the signal is low-, medium-, or high-power, C- or Ku-band, the basic parts of a satellite-to-home broadcasting reception system are the same. The major components fall into three categories: the dish and its mount, outside electronics, and inside electronics.

Dish and Mount. A satellite receiving dish is basically a reflector that gathers signals falling on it and focuses them into a narrow, concentrated beam. The larger the dish, the greater its ability to concentrate and, in effect, boost the satellite signal. Conversely, the stronger the signal, the smaller and less cumbersome the dish can be.

Dish size is also affected by the frequency of the signal transmitted. Given two signals of equal strength, the higher frequency signal can be picked up just as clearly on a smaller dish. Consequently, signals in the Ku-band (12 gHz) can be received adequately on a three-foot dish, whereas the same television program transmitted in the C-band (4 gHz) requires an eight- to twelve-foot dish for the same quality reception. The extra power and higher frequency of Ku-band DBS allow the use of dishes as small as eighteen to twenty-four inches.

Much as with a magnifying lense or mirror, the shape of the dish and regularity of its surface are critical to its effective operation. Most home dishes are circular at the edge and have a parabolic shape. The parabolic reflector bounces all signals falling on it to a single focal point in front of the dish. Dishes are fiberglass, solid metal, or perforated metal mesh and are shipped as a single unit or in sections ready for assembly.

The dish must be mounted with a clear view of the Clarke Belt—where the satellites are located 22,300 miles over the equator. The amount the dish must be tilted is its angle of elevation, which decreases as the dish is located further from the equator. For example, backyard dishes in Miami, Florida, have an elevation of about sixty degrees above the horizon; in Maine, only about forty degrees. Most dishes are mounted on a single metal support known as a polar mount that is aligned in a way that allows them to move freely from left to right from one satellite to another along the Clarke Belt. Motors allow for remote-controlled positioning of the dish from inside the house.

Outside Electronics. At the focal point in front of the dish where the signals are concentrated is the feedhorn, which directs the signals inside to the system's antenna. Because the wavelength of satellite signals is less than one centimeter, the antenna itself is shorter than one inch. From the antenna the signals enter the low noise amplifier (LNA), a small black box usually located in the feedhorn. Once boosted, the signals are shifted to lower frequencies by another black box, the downconverter, which may be bolted to the mounting pole or combined with the LNA. From the downconverter, signals enter the house through a cable that also carries the electrical power for the outside electronics and the dish positioning motors.

Inside Electronics. The cable is connected to a satellite receiver that further amplifies and processes the audio and video portions of the signal, to which it is tuned using the channel selector. From the amplifier the signal either passes directly to the television set or first to a descrambler in the case of subscription channels. The VideoCypher II descrambler, which has become the U.S. industry de facto standard, searches the satellite signal for an authorization code number transmitted by the program service.

Each decoder is assigned a unique code number. If its code is being transmitted, the VideoCypher II recognizes it, unscrambles the signal, and passes it to the TV set. This is known as an addressable descrambler or decoder. An addressable system allows subscribers to add or drop channels by simply ordering them over the phone or by mail. Addressability permits multiple program services to be accessed conveniently by a single decoder. If payment is not received, a program service removes the subscriber's access code from its transmission and the decoder will no longer descramble that programming.

Low-Power Satellite TV Outside the United States

Outside North America home satellite dishes are uncommon; there has been no international backyard dish phenomenon on the scale of the U.S. experience. In the Caribbean and Central and South America, private dish ownership is limited to hotel/resort operators or a few relatively wealthy individuals. A number of national and pan-European cable program services now are transmitted by satellite, but backyard dishes are not a common sight in Western Europe. This is partly due to restrictions on dish ownership in many countries and the uncertain status of low-power satellite-to-home broadcasting under national regulations and ITU rules. Although DBS has been recognized by the ITU, reception of signals from conventional C-band satellites by private individuals is still a gray area. Consequently, there is no standardized scrambling system for low-power services, and they are not marketed directly to individuals on a scale comparable to the U.S. model. The growth of cable television, the increasing availability of VCRs, and the cost of large dishes also pose difficulties for the development of widespread low-power satellite-to-home broadcasting in Europe in the near future.

European DBS

European advocates of DBS have had plans for high-power national and pan-European systems in development since national frequency assignments (up to five channels per country) were established by the ITU in 1977. Throughout the 1980s backers of government-sponsored and privately financed DBS systems had to overcome not only many of the same problems facing their U.S. counterparts but also the additional difficulties presented by Europe's multinational market. Among these are different technical standards for television, language preferences, varying national and local policies regarding licensing of receiving equipment and transborder broadcasting, and the need to establish multinational sales and collection systems.[6] The result has been a variety of DBS schemes, each with its unique aspects.

Europe's first DBS satellite was launched in November 1987 by an Ariane vehicle. The German-owned satellite, TV-Sat I, was to be followed within months by the identical French-owned TDF I. Each had four transponders powerful enough to reach a two-foot dish, costing about $1,200, with a mix of government and privately sponsored TV programming. Although TV-Sat I was abandoned after one of its solar panels failed to deploy, TDF-I was successfully launched in October 1988 and was expected to become fully operational in 1989.

Other countries with announced DBS plans in early 1989 included Britain, to launch a multichannel service (British Satellite Broadcasting) in late 1989 on a satellite called UK-sat; Sweden, Norway, and Finland, to begin Tele-X service on a jointly owned satellite in 1989; Italy and Britain, major partners with six other countries in a two-channel Olympus satellite, scheduled for 1989; and Ireland, with a system in early development stages. Japan, which has experimented with DBS since 1984, was proceeding with plans for a 1990 launch of a DBS satellite that would include at least one high-definition channel (see Chapter 8).

As was the case in the United States with USCI, European DBS plans are being challenged by at least one medium-power system. Luxembourg's

Ku-band Astra system, launched in December 1988, eventually will offer sixteen channels of multilingual commercial programming to subscribers using somewhat larger dishes, about four to six feet. Although all the other systems are intended to be limited in coverage area, the Astra system will be pan-European, with most of Western Europe in its signal footprint.

Summary: Three Forms of Satellite Broadcasting

It appears that three fairly distinct modes of satellite-to-home broadcasting will coexist in the near future. Low-power satellite-to-home broadcasting involves C-band satellites and eight- to twelve-foot dishes. It is primarily a U.S. phenomenon that began as eavesdropping on cable programming services, then evolved into an authorized commercial subscription service. DBS is a high-power service that requires expensive specialized Ku-band satellites with only five transponders but can reach two-foot dishes. It has faltered in the United States and is just getting off the ground in Europe and Japan. Both these forms face competition from future medium-power systems that can be operated using the newer generation of conventional Ku-band satellites and medium-size (three-foot) dishes and do not require DBS licenses.

It is reasonable to assume that eventually very small satellite receiving dishes will become common household appliances in many industrialized and some lesser-developed countries, as has the television itself, if only because receiver technology will continue to become more sensitive and less expensive and satellites more powerful. In Japan and Europe flat, checkerboard-size indoor receiving antennas have already been demonstrated that can pick up transmissions when simply aimed out the window.[7]

In the United States low-power satellite broadcasting has survived a difficult and controversial birth and DBS remains uncertain. In Europe a DBS satellite has been successfully orbited and others are scheduled. However, the fate of the various high- and medium-power services described above depends not only on the availability of satellites but also on the development of consumer demand and the solution of a complex international mix of political, economic, and technical problems. It may well be that Japan, with its powerful economy, leading-edge technology, and more unified telecommunications development policies, will ultimately be the first country to make high-power satellite-to-home broadcasting a common household convenience.

Notes

1. According to Mark Long and Jeffrey Keating, *The World of Satellite Television* (Mendocino, Calif.: Quantum, 1985), 18.
2. Title VI, Communications Act of 1934 as amended by the Cable Communication Policy Act of 1984, Public Law 98-549.
3. Eric Mankin, "Home Dishes Get New Signals," *Electronic Media* (1 January 1988): 34.
4. See J.L. Freeman, "FCC Scrambling Study Sees Marketplace Working," *Multichannel News* (16 February 1987): 5.
5. See "FCC Opens the Skies to DBS," *Broadcasting* (28 June 1982): 27.
6. For detailed discussion see "Direct Broadcast Satellites (DBS) and Communication Satellite Systems," in John Tydeman and Ellen Jakes Kelm, *New Media in Europe: Satellites, Cable, VCRs and Videotex* (London: McGraw-Hill, 1986), ch. 5, pp. 86–130.
7. See William Rabkin, "Flat Satellite Dishes Offer Alternative to Unsightly Dishes," *Electronic Media* (8 June 1987): 17.

PART III

The Expanding Video Horizon

6

The Revolutionary VCR

In only a few years the home videocassette recorder has become a common household appliance in the United States, Western Europe, Japan, and many other parts of the world. It has changed the way millions of people spend their leisure time with television. More significantly, by reducing viewers' dependence on television networks and movie theaters for video and film entertainment, the VCR is changing long-standing economic relationships among the electronic media.

Solving the Problem of Video on Tape

In the early days of television broadcasting a number of major electronics manufacturers sought to develop a practical method for recording television pictures and sound on videotape. Everything broadcast on television was either live or film. The only way for a viewer at home to preserve a live program was to film it directly from the television screen, a process known as kinescope recording. Audiotape recorders were introduced shortly after World War II and soon were in widespread use in the industry.

The recording of video on tape, however, was a much more complex challenge than audio-only. A television picture consists of thousands of times more electronic information than a simple audio signal. Any video recorder modeled after an audio recording system—a single, fixed recording head over which tape travels horizontally—would have to operate at such high speeds and with such large reels of tape that it would be impractical. For example, the BBC aired videotape broadcasts in 1955 using a system that required a five-foot reel of tape to record a thirty-minute program; tape speed was over sixteen feet per second.

The technical solution came from a California company, Ampex, that had been instrumental in the development of audio tape recorders in the 1940s. Ampex solved the recording speed problem by designing a four-section recording head that rotated vertically across the tape as the tape itself moved past the head. The tape moved at speeds similar to audiotape recording (about fifteen inches per second) because the spinning recording heads in effect multiplied the speed of the two-inch wide tape, producing a practical system for studio use.

The Ampex system was presented to the U.S. television networks in 1956 and was an instant success. Although it was as big as a kitchen stove and weighed much more, the Ampex video recorder was the progenitor of today's home VCRs and camcorders. By the late 1960s a more compact version of Ampex's record head system, called the helical scanning head, had been introduced. This and newly available microelectronic circuitry reduced the size and increased the reliability of recorders, making transportable recorders practical. Sony and the Dutch manufacturer Philips introduced videocassettes by 1970, eliminating the need to hand thread videotape. Sony's VCR system using three-quarter-inch-wide tape quickly became standard for industrial and commercial television use in the United States.

Debut of the Home Video Recorder

By the mid-1970s several consumer electronics companies were attempting to design a less expensive, smaller-format VCR that might appeal to home users. In 1975 Sony introduced the Betamax, featuring half-inch tape in a small cassette and a $1,300 price tag. Blank cassettes cost about $20 and recorded for only an hour. About the same time, Philips also introduced in Western Europe a domestic VCR that was incompatible with the Betamax. Although prices began to drop rapidly as sales picked up, the limited recording time of these early machines invited competition. Soon Sony's competitor, JVC, owned by the giant electronics manufacturer Matsushita, introduced the VHS format with its slightly larger cassettes and longer recording capacity.

By 1980 VCRs were appearing in homes in many industrialized countries, but penetration levels were less than 5 percent of homes with television. However, sales growth and competition between the rival formats continued to bring down prices and improve the technical quality of the recorders. Lower prices, the inclusion of built-in timers for automatic off-air recording, special features like slow motion and visual fast-forwarding, and increased recording times made the VCR even more attractive to the average consumer.

The VCR Becomes a Global Medium

In the early 1980s the growing availability of rental tapes and plummeting prices for VCRs and blank cassettes combined to double VCR penetration from 1982 to 1983 and then double it again from 1984 to 1985. Although gains were less spectacular after 1985, penetration climbed to 50 percent of U.S. television homes in 1988, with some major cities approaching 60 percent. In Western Europe VCR penetration in wealthier nations ranged from about 20 to over 40 percent, with the United Kingdom leading the way. In Canada, Australia, and Japan VCRs could be found in 40 to 50 percent of households. In poorer countries levels were closer to 10 percent. It is estimated that within the next few years VCR penetration will approach 50 percent of households in Europe and may go as high as 80 percent in the United States.

The VHS format now dominates VCR sales and cassette sales/rentals in the United States and in Europe where the original Philips format is a distant third behind VHS and Beta. In final recognition of the VHS format's conquest of the world market, Sony, the standard-bearer for the Beta format, announced in January 1988 that it would manufacture a line of VHS machines. The VCR is a global phenomenon, unavailable only where the standard of living or government policy prohibit it; and VHS is the international home video format.

The VCR and the Law: Home Taping

In the United States legal issues have played a key role in the growth of the VCR as technology and as an electronic medium. Perhaps the best known was addressed in the Betamax case. In 1976, not long after the first Betamax VCRs appeared, two major film and television program production companies jointly filed suit against Sony. Universal Studios and Disney Productions argued the use of VCRs to record their programs off television was a violation of their copyright and that Sony was selling machines used for an illegal activity.

Sony argued that recording and viewing television programs for home use was exempt from copyright law under the fair use concept. The U.S. Copyright Act of 1976 permits limited duplication of copyrighted materials by private individuals when such a use does not interfere with the copyright holder's rightful profits—the fair use concept. Limited copying of materials for educational purposes or by reviewers or critics are fair use examples cited in the law. The key legal question taken up by the courts was whether home videotaping—which was not mentioned in the Act—was also a fair use and could continue. Two lower federal courts split on the issue, first holding in favor of the home user and then against. Then in 1984 the U.S. Supreme Court held, in a narrow five-to-four ruling, that videotaping programs for later viewing was a fair use and exempt from copyright restrictions.[1]

Had the decision gone the other way, no one expected an invasion of homes and stores by copyright police confiscating VCRs and cassettes. But an anti-home recording decision would have given the studios the legal leverage to negotiate some sort of payment or tax from VCR and cassette manufacturers, perhaps a surcharge on each unit sold. As usual, this cost ultimately would have been paid by the consumer as an increase in the retail price of VCRs and tapes. The Betamax decision allowed manufacturers to continue producing machines and cassettes and setting prices without dealing with the major studios and other production companies. To the extent that contributed to the ever decreasing prices of these products, the decision helped to spur industry growth at a critical time.

The Home Video Industry Appears

Not long after the home VCR was introduced, many entrepreneurs recognized its potential as a distribution medium for films copied to videocassette. Some in major cities discovered a brisk market for the sale and rental of pornographic tapes. Others attempted to convince major motion picture studios to release feature films for home distribution on cassette. Among the first to succeed was a small company called Magnetic Video, which licensed fifty films from 20th Century–Fox in 1978, marking the beginning of the video sales and rental industry.

The home video industry (as it came to be known) started out as a little noticed mom and pop business in major cities where VCR sales were first blossoming; small shops carried 300 to 400 cassettes. But as VCR penetration exploded over the next five or six years, video stores popped up in malls and shopping centers everywhere. By 1988 about 40,000 video sale/rental outlets existed. Stores are now so common that competition is driving a trend toward consolidation. Larger stores owned by video rental chains are the industry's mainstay, and superstores have appeared that stock as many as 30,000 cassettes. A similar industry has taken root in Western Europe. Videocassette sale/rental stores are a familiar sight in urban areas in most of the world's industrialized countries.

In the United States the market for VCRs and tape sales/rentals also gave birth to a home video programming industry. At first these were independent companies, like Magnetic Video and Vestron Video (established in 1981), which negotiated the home video rights to films, duplicated the cassettes, and sold them to video stores. Soon, however, these independent suppliers were joined by others that had ties with major motion picture studios. For example, Magnetic Video was soon purchased by Fox. Columbia, Paramount, MGM/UA, and Disney all quickly established home video operations. By 1988 most of the largest suppliers of films to the home video industry were studio-owned, with the exception of Vestron.

Home Video and the Law: Cassette Rentals

Just as fair use was critical to the growth of VCR sales, another legal issue directly affected the growth of the prerecorded videocassette sales and rental industry. Once again the key issue was payments to production companies. When video stores first began renting movie tapes to customers, production companies (movie studios and others) were dismayed to learn that they could not require the video stores to pay them a share of the rental fees being collected.

Under the legal first-sale doctrine, a studio's right to any proceeds from a movie cassette ends with its sale to a video store or a distributor. That is, when a video store or distributor purchases a cassette, it is also paying for the right to do with it as it pleases—to resell it at a profit or to rent it out repeatedly until a profit is made. Retailers generally have the right to resell or rent out as they please the stock they purchase from wholesalers or distributors, unless they have agreed in advance to some other arrangement.

Unable to successfully challenge the first-sale doctrine in the courts, the production companies tried to make up for the lost rental revenues by charging high prices for movie cassettes. Although $70 and $80 cassettes produced income, it soon became apparent that the high prices were also causing most potential purchasers to be renters, filling the pockets of rental store owners but not the studios. A breakthrough came in 1984 when Paramount began selling some popular films on cassette for $24.95. Other studios, including Disney, soon followed with similar reductions on many titles. Thus, the first-sale doctrine has contributed to the growth and profitability of the video rental business by protecting operators' rental revenues and indirectly reducing the prices paid by store owners and their customers for cassettes.

A Smaller, More Elusive TV Audience

In the United States VCR penetration has reached what some marketers call critical mass—the point at which so many are used every night that fundamental changes are taking place in the nation's viewing habits. The most basic change has been a shift in audience away from the three major television networks. By 1980 the rise of cable and independent local broadcasting stations had begun eroding the audience share of the three major television networks. The arrival of the VCR compounded this network audience erosion by making alternative programming all the more accessible. The VCR also provided an alternative of its own: rented movies. Together, movie rentals, cable, and independent stations have reduced the three-network prime-time audience share from around 90 percent to around 75 percent on a typical evening.[2] Nonnetwork programming has been affected also. For example, some stations report ratings drops in childrens' shows, which they suspect is attributable to VCR use after school.[3]

Although it has been predicted that in a few years Americans will spend as much as one-fourth of their TV viewing time watching prerecorded cassettes,[4] the primary use of the VCR today is time shifting—recording shows for later viewing at a more convenient time. A 1987 Nielsen survey reported 49 percent of taping was done with the set off and 18 percent with the set on but tuned to a different channel.[5] In both cases the scheduled time of the program being taped is, in effect, shifted by the viewer because it conflicts with another program or because the viewer cannot watch then.

The balance of taping (33 percent) was done with the set on and tuned to the same channel being watched, presumably so the show can be watched two or more times. The fact that most home taping is done to watch programs that otherwise would be missed was an important consideration in the majority opinion in the Betamax case.

Most time shifting doesn't take audience away from the networks; over 70 percent of television programming recorded on VCRs is network fare.[6] However, it does make the audience much more elusive. Traditional measurement methods for audience ratings have been geared to counting viewers who are watching programs in real time and not recording shows for later viewing. Although the major national rating services have recently switched to more sophisticated measurement devices and some use of VCR diaries, comprehensive national data on time-shifted program viewing is still not available. In local markets, for the most part, viewers who time shift a station's programs simply disappear and the rating drops.

Particularly affected by the VCR has been weekend programming. An NBC network study revealed that over half of VCR playback occurs on Friday, Saturday, and Sunday nights (16, 20, and 18 percent, respectively).[7] Although most of what is being played back is network fare, VCR use is hurting the ratings of the regularly scheduled programs.

Throughout the viewing week VCR owners appear to be skimming from the best of the network programming by taping their favorites and watching them at more convenient times. This disrupts what the networks call audience flow—the tendency of viewers to stick with a given network throughout the evening because it carries a favorite program. Consequently, new or less popular programs scheduled next to hits no longer get as much ratings help as they once did.

In addition to audience flow and ratings problems, the VCR causes network and advertising agency concern because of its ability to eliminate commercials when viewing recorded programs. Many VCR owners fast-forward through commercials when they view recorded programs. The first major British study of VCR use found that 74 percent of owners with remote controls skipped commercials during playback.[8]

Although it is unlikely that every commercial is victimized by the *FFwd* button, the appeal of zipping is obvious, especially in U.S. commercial programs. With nearly twenty minutes of commercials in a typical hour of prime-time network programming, zipping past them makes it possible to watch a one-hour show in forty minutes, a two-hour movie in eighty minutes. In addition to time shifting, the ability to watch a program with fewer or no commercials, in significantly less time, must be a major reason many VCR owners record network programs.

The VCR's impact on television viewing is reportedly causing changes in network programming strategies. For example, fewer theatrical films are shown in prime time because most have already been rented by VCR owners. Flashier, more sensational, more "adult" network programming has appeared to counter VCR use, especially on Friday and Saturday nights. Programs have been developed to appeal to older audiences, who are less likely to own VCRs. Much effort has been expended to make commercials more attention-getting and appealing, so they won't be zipped on replay. In addition to trying to beat the VCR, networks and local stations have begun packaging home video versions of some popular programs and miniseries. Many have sold well in the United States and are popular with European audiences.

The VCR Tail Wags the Film Industry Dog

In the early 1980s film production companies—both the major Hollywood studios and independents—made their money by exhibiting films first in theaters, then on pay cable (like HBO), then on the commercial television networks. Generally, films that did well in theaters could command high fees from pay cable and network TV. Home video was nothing more than a sideline. It was difficult to foresee that before the end of the decade home video would become Hollywood's biggest source of revenue.

Today major motion picture studios make more profit from cassette sales than from theatrical exhibition. The American public spends as much money on cassette sales and rentals as on theater tickets. The video store, which used to reside at the lowest level of the film distribution hierarchy, now is second only to theaters in order of access to new films, which hit their shelves within months after theatrical release.

Video rights to a major picture can bring its studio $10 million or more—enough to cover half the cost of production in many cases. Films that lose money at the box office can sometimes turn a profit from cassette sales. Consequently, the VCR, by generating home video revenues, has turned what were filmmaking companies into video publishers as well. Films are financed and produced in anticipation of home video revenues. Many classic films have again become income producers, after being dusted off and put on the video dealer's shelf.

Profits from sales and the inability to make money from rentals have led video publishers to the doorsteps of the country's major mass merchandisers with their wares. Discount chains, bookstores, record shops, and even direct mail are combined with reduced prices in a sell-through strategy designed to encourage the purchase of cassettes. One of the first major sell-through successes was Paramount's *Top Gun.* It sold almost 3 million copies with a $26.95 price tag made possible in part by Pepsi commercials added to the tape; a first for a major film on cassette.[9]

Ironically, although the VCR has generated new revenue streams for the film studios, it has also reduced their control over their product. Before home video rentals a studio had complete control over nearly every showing of a film. It knew when it would run in theaters, on pay cable channels, and on network TV; and every one of those bookings produced accountable revenue. Now rented copies of *Top Gun* or classics like *Snow White* are being seen thousands, perhaps millions, of times the world over without studio knowledge or profit.

Perhaps worse from the studio's standpoint, many titles are being illegally copied and sold in the United States and elsewhere. The solution to the piracy problem is an effective electronic coding system that allows a tape to be viewed but not copied. Although several studios have announced plans to use one such system, called Macrovision, technical problems and expenses have delayed widespread encoding of tapes. Of course, even if all tapes were encoded beginning today, it takes only one clever pirate to defeat the system and contaminate the marketplace with uncoded cassettes of a film, which in turn can be copied by other pirates.

Conclusion: Weighing Gains and Losses

The spread of the VCR across most of the industrialized world in only a few years has opened an entirely new electronic pathway for the flow of entertainment and information programming within and among nations. It has brought within the reach of the average person thousands of movies, old and new; music videos, popular and classical; instructional and documentary tapes; even exercise programs. The VCR's positive potential for promulgating the visual arts and helping to improve the quality of life seems enormous. On the other hand, the VCR has an equally great negative potential for extending the influence of socially detrimental types of programming.

Like radio and TV before it, the VCR's development will reflect economic and political realities rather than wishful thinking. The VCR is already seen as a cultural and economic threat in some instances. Countries fearing cultural invasion via VCR are already taking steps to control its impact. The French government, for example, has outlawed video distribution of any film until a year after theatrical release to protect the national film industry. Taxes on foreign VCRs and cassettes are used in France and other countries to protect indigenous electronics industries and cultures.

The VCR's impact on existing electronic media may also conflict with long-range public-interest goals of telecommunication policymakers in the United States and elsewhere. For example, in Europe, where most national cable systems are still in early stages, there is concern that cable's slow development is partly caused by the rapid spread of the VCR as an alternative medium. In the United States the extent to which the VCR may weaken the television networks may ultimately be reflected in a weakening of their local affiliates, perhaps the last major remnants of localism in television broadcasting. As is the case with any revolution, the VCR's dramatic arrival as a new electronic medium may produce numerous interesting, perhaps unforeseen, results.

Notes

1. Sony Corporation of America v. Universal City Studios, 457 US 1116 (1984).
2. See, for example, "Ratings down 10%, Shares off Four Points," *Broadcasting* (9 November 1987): 35.
3. Richard Mahler, "VCRs Vs. TV Stations: Broadcasters Fighting Home Video Onslaught," *Electronic Media* (17 August 1987): 1.
4. Ibid., 30.
5. "Facts and Figures," *Multichannel News* (27 July 1987): 42.
6. Ibid.
7. Richard Mahler, "VCRs Alter Viewing Habits," *Electronic Media* (17 August 1987): 30.
8. Laurel Wentz, "British Study Shows Most Zap Ads," *Electronic Media* (21 March 1985): 18.
9. See Richard Zacks, "Cassettes Rewrite Studio's Books," *Channels* (December 1986): 90.

7
Laser-Optical Disc Systems

To many the term *laserdisc* brings to mind the compact audio disc, or CD, which is rapidly displacing the vinyl record as the medium of choice for recorded music. However, the audio CD is only one way laser-optical technology is used for storing and retrieving information. This chapter reviews the development of optical disc technology and its growing importance as a new electronic medium.

Laser Fundamentals

The light we are most familiar with consists of a jumble of light waves of various frequencies that bounce irradically off objects they strike. A beam of this natural light also tends to spread out as it radiates. This is one reason that a small-diameter flashlight beam illumines a large area at a distance. Despite the use of lenses and reflectors to focus and control it, the incoherency of natural light limits its usefulness.

A laser is a device that emits a beam of very unnatural light – unnatural because it is coherent. Developed in the early 1960s, lasers come in many varieties but operate on the same basic principles. Light or electricity is used to energize certain liquids, gases, or solid materials in a closed container. The energized material emits coherent light waves that are amplified and allowed to pass out of the container. The laser derives its name from this process: light amplification by stimulated emission of radiation.

The coherent light produced by a laser consists of a single frequency and is highly directional. It can be focused into an extremely narrow beam that does not spread out. Laser beams pass through transparent material harmlessly but can be used to heat or even melt material that absorbs light. The laser has become a remarkably useful device with applications in scientific research, manufacturing, medicine, and communications. Lasers cut through steel, survey land, etch microscopic computer chips, remove tattoos, and repair damaged retinas. Laser light pulses can also transmit information that has been converted to digital code. This digital format is used to store information on a laser disc.

Optical Disc Fundamentals

Imagine a microscopic pit—a depression so small that 1,200 would fit side by side across the period at the end of this sentence. Strings of billions of pits this size form concentric circles on the surface of an optical disc and store the information recorded on it. The information takes the form of computer data—the series of 0's and 1's referred to as binary code. Information that is not already in this form, such as video and audio signals, must be converted. This is accomplished through a process called sampling in which the electronic signal representing the sound or pictures is read or sampled thousands of times per second. These values are stored as a long string of binary numbers in computer memory. A laser is used to burn this code into the face of a blank master disc. Each 1 is represented by a pit; the 0's are the blank spaces in between.

The master disc serves as a template from which thousands of others are pressed in a manufacturing process similar to conventional record pressing but much more exacting because of the microscopic size of the impressions being made. The completed discs are made of tough plastic material covered with a highly polished metallic surface. This surface, which contains the coded information, is in turn protected by an outer layer of clear plastic.

During playback, a very low-power laser, no bigger than a pencil eraser, is focused on the continuous track of pits on the disc surface. The beam of laser light is about the same width as the pits themselves. As the disc rotates beneath the laser, the beam either strikes a pit and is scattered, or strikes the smooth, mirror-like surface of the disc in between the pits and is reflected back toward the laser. As the disc whirls past the laser at over 1,000 RPM, a photosensitive cell detects the series of reflected light pulses and converts it into a digital electrical signal. If the system is being used for computer information storage, the signal remains in digital form. On a video or audio player, however, the digital code is converted into an analog signal that can be reproduced as pictures on a television screen or sound through a speaker.

Optical discs provide a number of advantages as an information storage medium. They are durable; the clear plastic coating on the disc protects the reflecting surface from fingerprints and small scratches and is washable. Furthermore, because the information on the disc is read by a beam of light that does not cause wear to the surface, optical discs do not wear out like magnetic discs and tape or vinyl records, in which the pickup head touches the surface, eventually wearing it down. Because the laser head does not touch the disc, it can move rapidly to access information on any part of the disc or read the same track over repeatedly without damage. Because the laser disc stores information in a much smaller physical space than tape (audio or video), it can access any part of it nearly instantaneously.

Most important, because the information is encoded digitally, it can be reproduced much more precisely. Error detection circuits built into the player "guess" the value of a damaged or missing bit of code. Random information like noise can be recognized and eliminated before transmission. The result is audio and video reproduction that is clearly superior to any previous method. Regardless of the type of information stored, the laser optical disc can pack more of it into a smaller space and in a more durable and accessible format than any electronic medium yet developed, other than the computer chip itself. Consequently the laserdisc has emerged in several forms as an important electronic communication medium.

Videodisc: Philips and RCA Systems

In 1979, after the VCR had begun to appear in American, West European, and Japanese households, the first laser videodisc players were brought on the U.S. market. They were developed by Philips, the giant Dutch electronics company, and manufactured by Magnavox and Pioneer under the name Laservision. Videodisc manufacturing soon began in single plants in the United States and Japan. Philips and Pioneer introduced similar players in Western Europe in 1982. Prices were $700 to $800.

About this same time RCA introduced an incompatible videodisc system in America and Europe under the brand Selectavision. It featured less sophisticated technology in a less expensive machine ($500). Unlike the laser optical system, the RCA's player used a stylus traveling in a groove in the disc surface. In the bottom of the groove were slots comparable to the pits in a laserdisc. As the stylus passed over each slot it registered differences in electrical capacitance from one slot to another, picking up the digitally encoded picture and sound signals. While the laserdisc was an entirely new method of information storage, the CED (capacitance electronic disc) system was more like a hybrid of digital information storage and conventional phonograph technology.

By 1983 it was clear that both players were generating disappointing sales. The market was split by the two different player formats and offered few movies on videodisc in either. Additionally, both videodisc players entered the consumer electronics market after the VCR had already established a small but significant beach head. Given a choice between a high-tech video playback machine with little software for sale and virtually none for rent and a lower-tech machine with recording capability and more software, consumers in America and Europe chose the VCR. While VCR sales climbed rapidly into the millions, videodisc player sales never took off.

After selling fewer than 100,000 laserdisc players, Philips ceased production in 1982. RCA cut prices on its players to $300 and less, attempting to make its profit on the sale of the discs rather than the players. Although it eventually sold half a million CED players, RCA stopped production worldwide in 1984, unable to compete with the growing popularity of the VCR and the expanding libraries of movies on cassette.[1] Philips' and then RCA's pullout appeared to mark the demise of the videodisc as a major home entertainment medium. However, Pioneer, Philips' manufacturing partner, has continued to sell Laservision players and press laserdiscs, keeping the videodisc alive as an entertainment option for high-tech enthusiasts and pursuing its industrial applications beyond the consumer entertainment market.

Compact Disc—Audio

The laserdisc was overwhelmed by the VCR in its bid to become a home video entertainment medium, but it has enjoyed dramatic acceptance as an audio medium. One of the selling features of the laser videodisc was the high-quality digital sound it produced along with the video. In 1983 and 1984, just as the videodisc was being declared dead, Philips, Sony, Pioneer, and other manufacturers were discovering an enthusiastic market for audio-only laserdisc players. Only about five inches in diameter, the audio laser disc, or compact disc, can hold over an hour of remarkably noise-free digitally recorded sound. It operates exactly the same way as the videodisc and has the same advantages of durability and rapid access to any portion of the recording.

By 1988 CD sales had mushroomed from hundreds of thousands to hundreds of millions per year in the United States, Japan, and Western Europe. About 40 percent of the market was in the United States, where some 50 million CDs were sold in 1986 and 100 million in 1987. The familiar black vinyl LP, which had already taken a back seat in sales to the audio cassette in many record stores, lost even more ground as record companies converted their catalogs to CD. With prices dropping to $150 or less for no-frills models, CD players could be found in nearly 10 percent of households in the United States, the United Kingdom, the Netherlands, and Japan.

Unlike the videodisc, which suffered from two incompatible formats, the CD enjoys a uniform international standard, as do LPs and audiocassettes. It is similarly versatile and lends itself to a variety of music packaging and pricing arrangements. While most CDs are long-play, containing the equivalent of an entire LP album or more, a short play or CD-single version is also available. This three-inch disc holds up to twenty minutes of sound and can be played on a conventional CD player with a plastic adapter, reminiscent of the snap-in adapters once used with 45-RPM vinyl records. These mini-CDs sell for as little as three dollars and are aimed at the growing number of portable CD players finding their way onto the market.

Compact Disc—Video

The success of the audio CD has given new life to the original concept of the videodisc, in a somewhat different form. Taking their cue from the success of music videos on television and on videocassette, U.S. and European manufacturers unveiled in 1987 the video compact disc, or CDV. The CDV offers about twenty minutes of audio and five minutes of video for about eight dollars. It is played on a CDV player that can take both CDVs and regular CDs. Though CDV manufacturers made enthusiastic predictions about its future, 1988 was still too early to tell whether the CDV would share any of the success enjoyed by its high-flying audio-only counterpart.

Given that the original twelve-inch laser videodisc has now been joined by three- and five-inch CDs and five-inch CDVs, combination players have appeared that can handle all of these formats interchangeably. These "combi" players can become the center of a laserdisc home entertainment system and are symbolic of the optical disc's versatility as a video and audio information storage medium. However, audio and video playback are only the beginning.

The Interactive Laserdisc

When Philips first developed optical disc technology in the late 1970s, it was apparent that the system's ability to rapidly access information on any part of the disc had potential applications well beyond the home videodisc player. During the 1980s, while consumer versions of the technology had their ups and downs in the marketplace, steady progress was being made in developing interactive optical disc systems for educational and industrial applications.

In a sense, any laserdisc system that will respond to simple selection commands from the user is interactive. However, the term is usually reserved for more sophisticated systems in which the program viewed is designed to allow the user to follow an individualized, self-paced sequence through it. The optical disc's ability to randomly access any portion of recorded information in an infinite combination of sequences and to remain on a single frame indefinitely makes it ideal for such applications.

One of the first demonstrations of this capability was the Aspen Movie Map, developed at the Massachusetts Institute of Technology in 1977. A laserdisc was programmed with pictures of the streets and historic buildings in Aspen, Colorado. Using a computer linked to the disc player, viewers could take a self-guided tour of the town, free to turn right or left or continue ahead at each intersection. Passing by an historical building, a viewer could take an inside tour by pushing an appropriate button.

Today a wide variety of interactive systems are in use, ranging from simple stand-alone laserdisc players to sophisticated systems in which several players are linked by a controlling computer that can add its own graphics and other materials to the output. The interactivity of the system is limited only by the amount of video stored on discs and the sophistication of the controlling computer. Consequently, interactive videodisc systems are being used increasingly in private industry, government, and education for a wide variety of applications. These uses fall into two broad categories: training or education and consumer information/sales.

Schools, businesses, and government agencies all have found training or educational uses for interactive disc systems. Universities have produced interactive programs covering many traditional subjects at the secondary and college levels, from chemistry to gymnastics. Government agencies and businesses have developed interactive training systems covering everything from auto repair and inventory control to simulations of high-tech equipment operation. Self-contained systems are commonly set up in kiosks in malls and other public places to provide information to shoppers and passersby. Kiosks can now be found in some stores that allow customers to rapidly view, and even order from, entire inventories or catalogs of merchandise not on the showroom floor.

The public's increasing contact with interactive disc systems on the job, at school, or in the mall, combined with widespread enthusiasm for the audio CD as a home appliance, has encouraged the marketing of five-inch interactive compact disc systems for the home, referred to as CD-I. Interactive games and how-to programs like home repairs or guitar lessons are in various stages of development. CD-I discs can also convert a database, like an encyclopedia, into an interactive audio-video learning tool. For example, Grolier announced a CD-I version of its twenty-volume encyclopedia that will include short video clips, audio, and even games in addition to 10 million words of text. CD-I discs will be played on the family CDV player using a joystick or keyboard similar to those now used for conventional video games.[2] Such electronic encyclopedias and games, however, only begin to suggest the optical disc's enormous capability for information storage.

Database Discs: CD-ROM

Popular music and video on optical discs seem to be the hook needed to lure consumers into buying CD systems in various forms. However, a single five-inch CD that can hold only a few minutes of video or an hour of music can contain the equivalent of 300,000 pages of text or data. Larger discs can hold millions. This extraordinary capacity for storing large amounts of text and data in very small spaces has at least two important applications: increasing computer memory and reducing the amount of space required to store documents of all kinds.

Attached to a personal computer, an optical disc can provide greatly enlarged memory or data storage capacity. One five-inch CD holds as much information as about one thousand conventional five-inch floppy disks. An optical disc used this way is called a CD-ROM (CD read-only memory). The players are referred to as drives, the same as conventional computer disk drives.

CD-ROMs make possible convenient access to large prerecorded databases: mathematical tables, census data, encyclopedias, telephone directories, catalogs, manuals, periodical indexes, and similar bulky collections of text or data. In 1988 several major U.S. computer manufacturers introduced CD-ROM drives (players) that could be connected to their personal computers. Designed to accommodate standard five-inch CDs, the drives sold for about $1,000. A noteworthy example of the versatility of the laserdisc, the Apple Computer drive featured headphones and speaker connections allowing the operator to enjoy music CDs when not using a database disk in the drive.[3]

As was the case with the video laserdisc a decade earlier, the CD-ROM will not generate widespread interest unless a variety of software is developed for it. The first CD-ROM databases were either custom-made or aimed at fairly specialized markets. For example, large libraries already have CD-ROM versions of encyclopedias and periodical indexes available for patrons to use on personal computers or dedicated CD-ROM terminals. By 1988 CD-ROMs aimed at somewhat broader markets were becoming available. Examples included *Small Business Consultant,* a compilation of over 200 government publications ($149); *Stat Pak,* a collection of U.S. government economic and political facts and figures ($125); and *Bookshelf,* a selection of reference materials for writers ($295).[4] As software becomes available and drive prices come down, the CD-ROM is expected eventually to replace many of the library materials now kept on microfilm and microfiche and provide personal computer users with libraries of information at their fingertips.

Businesses and government agencies with a need to store and access large volumes of paperwork are taking advantage of the optical disc's data capacity by converting paper records to computer-readable form and transferring the data to optical discs. For example, one insurance company was faced each year with finding 39,000 additional square feet of office space to house nearly 2 million new file folders containing policies and other customer records. It announced plans to convert to an optical disc storage system that would store their previous seven years' accumulation of documents in only 100 square feet of space.[5]

Erasable Optical Discs

Although the optical disc is a far superior information storage medium, it has not replaced the familiar floppy (magnetic disk) in computer applications or the VCR or audiocassette recorder in the home or at work because of the lack of an economical system that will record as well as playback. The pits on a conventional optical disc are permanent; once a disc has been pressed it cannot be erased, nor can new data be added. Because of the potential market for an erasable system, several major electronics firms worldwide have invested heavily in the development of new techniques for storing information on optical discs.

An early breakthrough was announced by Matsushita of Japan in 1983. In this pitless system the disc's smooth surface is made of an alloy that changes its molecular structure when struck by certain types of laser beams. Where the recording laser strikes the surface, the material crystallizes and becomes reflective. A second laser can erase information because it causes the surface to decrystallize and lose its reflectivity in the spots it strikes. Although a technological breakthrough, the Matsushita erasable optical disc systems originally cost $70,000 to $100,000 each and have been purchased by only a few large corporations and agencies.[6]

Although cheap, erasable laserdiscs and affordable disc recorder/players do not appear to be imminent, a record-only version of the laserdisc has been developed. Known as WORM systems (for "write once, read many times"), they are finding increasing being used for document and data storage. In this case, a recording laser beam in the disc player or drive permanently alters the surface of the disc, which cannot be erased. In applications like document storage this permanence and durability is an advantage.

Even in applications that do not require permanent records, a WORM system can be cheap enough and hold so much information that it is an economical alternative to erasable magnetic disks. In other words, if WORM systems can be made cheaply enough, it appears that for many users their permanence may not be a problem. With hundreds of times more storage capacity than a similar-size magnetic disk, WORM discs, in effect, might allow users to consider disc storage space disposable.

In home applications, however, the economics are different. With video- and audiocassettes cheap and readily available, consumer versions of erasable laserdiscs and disc recorders will need to be priced competitively, regardless of their superior reproduction capabilities. Consequently, Matsushita, Sony, Philips, and others continue the search for an economic erasable optical disc system that might one day become an affordable office or household appliance.

Notes

1. For an interesting postmortem of RCS's failed videodisc efforts, see Margaret B.W. Graham, *RCA and the VideoDisc: The Business of Research* (Cambridge: Cambridge University Press, 1986).
2. Bob Young, "CD, CDV and CD-I: Futures Unlimited," *Billboard* (26 September 1987): C-3.
3. Patricia Zengerle, "Apple Endorses CD-ROM in Drive Debut," *Multichannel News* (7 March 1988): 30.
4. Patricia Zengerle, "CD-ROM Is in Bloom as Products Debut," *Multichannel News* (7 March 1988): 1.
5. Marsha Johnston Fisher, "Big Blue Moves to Image Processing," *MIS Week* (2 February 1988): 1.
6. "Matsushita Develops Erasable, Optical Disc," *Multichannel News* (11 April 1983): 58.

8
High-Definition Video

In 1981 Japan introduced the television world to its future: a new generation of receivers and studio equipment that produced pictures so sharp they looked more like film than TV. Japan, Western Europe, and the United States are now involved in an economic and technological struggle that may determine the new world standard for high-definition television. At stake is not only the multibillion dollar international TV equipment market but also the fate of existing television broadcasting systems and manufacturers worldwide.

The Development of Modern Television

Various theoretical proposals for television systems predate the twentieth century, and crude mechanical scanning systems were conceived in the 1880s. Although experiments with mechanical scanning systems continued, the modern electronic television era began in the 1920s with the development of the electronic picture tube in the United States and other countries. By the 1930s laboratory experiments had become experimental program transmissions to small numbers of receivers in the United States, Western Europe, and the Soviet Union. By the late 1930s TV technology had developed to the point that its use as a mass medium seemed feasible. Mass production of transmitters and home receivers required manufacturers and government to establish the technical particulars of such a system—that is, a television standard.

The first country to adopt a technical standard for television was Great Britain in 1937. The United States settled on a different, slightly more advanced standard in 1941. That year the National Television Standards Committee, an industry advisory group, proposed, and the FCC accepted, the NTSC system found in homes today. The NTSC standard was the high-definition TV of its day, the product of twenty years of engineering effort to enhance the first shadowy, grainy laboratory TV pictures.

Affordable receivers and the growth of the television industry were delayed by World War II. In the 1950s the television industry mushroomed in the United States, firmly establishing the NTSC standard. European industry growth, however, was further delayed by postwar economic recovery. That delay allowed a second European standard somewhat better than NTSC to be introduced in the 1960s. Although color transmission, stereo sound, and other refinements have been added, contemporary TV systems throughout the world are essentially enhancements of these early standards.[1]

Lines, Frames, and Resolution

The technical aspects of television transmission and reception are complex and not of great interest to most of us as long as we can push a button and watch a program. However, a few basic technical concepts are essential to understanding the vocabulary of the growing international controversy over high-definition TV.

If you could freeze the picture on your TV set and examine it closely, you would see that it is made up of horizontal lines of dots. Each dot, or picture element (called a pixel), is a tiny bit of phosphorescent material painted on the inside of the picture tube that glows when struck by a scanning electron beam. The dots are arranged in lines across the screen from one side to the other. The beam scans the dots, one line after another, "lighting up" a mosaic that duplicates the signal it has received—the television picture.

Generally, the more lines the screen has, the sharper and clearer the picture will seem. This apparent sharpness or clearness is referred to as resolution or definition. Today's 525-line NTSC standard was accepted in 1941 as a reasonable tradeoff between resolution and expense, as was the 625-line standard adopted later in Europe.

A moving TV picture is actually transmitted as a series of still pictures or frames, much like the frames that make up a length of movie film. In TV and in film, the apparent motion we see is caused by the speed at which the frames are transmitted; it is fast enough to fool the eye. Early film pioneers discovered that a projection speed of twenty-four frames per second (FPS) was fast enough to produce smooth, unblurred motion on the screen. Today 24 FPS is still the international standard for film.

Television systems were designed to project frames at comparable speeds. To simplify receiver design and reduce expense, speeds were chosen that were compatible with the speed of the household electrical current the television receivers would use. In the United States, which uses 60 Hz current, a 30 FPS system was developed. In Europe 25 FPS was chosen, compatible with the 50 Hz electrical system.

The World's Incompatible Television Systems

Today the 525/30 NTSC system is used predominantly in the Western Hemisphere and Japan. The 625/25 system, known as the CCIR system after the international committee that adopted it, is used in most Western European countries. China, the USSR, and Eastern European countries use a 625/25 system incompatible with CCIR. Other countries have chosen from these three major systems or variations of them, often considering politics or program availability before technical factors.

Receivers, transmitters, and studio equipment using one standard are completely incompatible with the other TV systems. Thus, there is no uniform world standard for television technology. Conversion equipment must be used, for example, to play back a British videocassette on an American TV or to allow an American network to simultaneously broadcast a program being aired live on a French network.

Because of the considerable investment they represent on the part of the government, the television industry, and the consumer, television standards tend to be self-perpetuating. Changing from one standard to another means rendering obsolete all present video production and transmission equipment, including the TV sets owned by viewers. Improvements in existing systems have, up to now, been adopted only when they were compatible with the technology in place.

National pride and income are also factors. Enough prestige is attached to television as an international medium that several countries still contend they were the first to invent it. The United States, Japan,

Western European countries, and others compete for positions of world leadership in electronics manufacturing. Setting the standard and controlling the patents for a major consumer electronics product like television receivers confer prestige and revenue. Consequently, until 1981 there was little talk of upgrading all the world's TV systems to some sort of unified, superior standard, although it might be technologically feasible.

Japan Unveils Its High-Definition System

At a 1981 San Francisco meeting of the Society of Motion Picture and Television Engineers, a consortium of Japanese broadcasting and electronics companies demonstrated a prototype television system that generated unprecedented interest and enthusiasm. At a later Washington, D.C., demonstration, normally staid television engineers, broadcasters, and regulators called the Japanese system "more than 100 percent better" than present systems, "fantastic," "extraordinary," "sensational," "almost three dimensional," and "a truly unique experience in television viewing."[2] It was called the NHK system after the Japanese public broadcasting corporation. NHK helped organize and fund the consortium's research, which began in 1968.

The NHK-HDTV system features a screen shaped more like a movie theater screen or a 35 mm photo print than a conventional TV set. Conventional TV screens have an aspect ratio of 4:3 (four units of width for each three of heighth). The HDTV screen aspect ratio is about 16:9, almost twice as wide as it is high. The extra width allows television viewing of theatrical films without the need to clip the edges, as is now done in TV broadcasts.

But the picture itself, not its shape, has caused the most excitement. It consists of 1,125 lines, more than twice the NTSC standard and nearly double the CCIR. The 1,125/30 system produces a picture with a resolution approximately that of projected 35 mm film. Even as close as three feet it is difficult to see any lines in the screen. As a result it displays none of the usual flaws of conventional TV pictures and is particularly effective when used in a large-screen projection television. The receiver is the foundation of NHK's extensive HDTV system, including cameras, videotape recorders, switchers, editors, and transmission equipment.

The HDTV Compatibility Problem

Because of the additional lines and wider screen width, the NHK system is completely incompatible with any of the world's existing televisions.

Moreover, the larger, more detailed picture contains about five times as much electronic information as a conventional picture. To broadcast this signal, NHK developed a special transmission system, called the MUSE system, which requires television channels with greater signal capacity than those already in use by stations around the world. In short, to make possible terrestrial HDTV broadcasting using the Japanese NHK system, not only would all equipment have to be replaced from the studio camera to the home receiver, but the channel assignments for TV stations would have to be completely restructured. The necessary increase in channel capacity could be accommodated only by making fewer channels available or taking spectrum space away from other telecommunication services. Although converters might be manufactured to allow existing sets to receive HDTV transmissions in conventional format, the TV channel capacity problem has not been solved.

The U.S. and European Responses

Despite compatibility problems, Sony and other manufacturers have actively marketed the NHK system worldwide and pressed forward with efforts to make it a new world standard for television. The result has been an international video standards controversy.

In the United States, where most color televisions are already imported from Japan, Korea, and their competitors, the major concern is not the impact on TV receiver manufacturing, but the possibility that HDTV would make all current TV station and network transmission and production equipment obsolete. Broadcasters especially are concerned that HDTV distributed by cable, by satellite, or on tape or disc might create its own audience, who will buy HDTV sets. This would leave local TV stations and their networks the lowest resolution video distribution medium in town—the video version of AM radio.

Various U.S. groups have proposed alternative schemes for HDTV systems that would be compatible with the existing television system. NBC has proposed a 1,050-line system that could use existing TV channels and be viewed on 525-line receivers.[3] Called the advanced compatible television system (ACTV), it would give viewers the option of keeping their old sets and watching the low-definition version of a program, or buying an ACTV set and seeing the same show in higher definition and wide screen, much the way compatible color transmission can be seen on black and white receivers. However, ACTV was still in the computer simulation stages in 1989, and it remained to be seen how it would compare with the NHK system under actual broadcasting conditions.

In Europe the NHK system is even less compatible than in the United States because of its 30 FPS speed. Furthermore the economic impact of a Japanese HDTV invasion would be considerably greater in Europe than in the United States. Not only would European broadcasters face the same obsolescence problems as their U.S. counterparts, but European manufacturers would lose much more. Most color receivers sold in Western Europe are made there. Having lost out to a Japanese VCR system (VHS), Philips and other major European manufacturers are determined to develop an HDTV system that can be manufactured and introduced under European control.

A consortium of major European electronics manufacturers, broadcasters, and government agencies has developed a 1,250/25 system geared to their present 625/25 standard. The Eureka system is incompatible with existing broadcasting equipment and probably will be used initially for satellite-to-home broadcasting. Before converting to 1,250-line transmission, European countries have plans to phase in an improved version of the 625-line standard. The Eureka HDTV system serves as a rallying point for European manufacturers concerned that their Japanese competitors might eventually invade their home markets with NHK-HDTV if no European alternative exists.[4]

Both in Europe and the United States other HDTV schemes have been proposed; all but NBC's require more than one channel, and many require decoders for reception on regular TV sets. In early 1989 all but the NHK system were still in the prototype stage. While electronics companies, broadcasters, and regulatory agencies in the United States and Europe responded to the HDTV challenge, Japanese companies proceeded with plans to produce and market HDTV receivers, VCRs, disc players, and studio production equipment, with or without a unified world standard. In Japan HDTV coverage of the Seoul Olympics was viewed by potential purchasers in 200 public locations.[5]

HDTV Studio Production Efforts

While the issue of international HDTV broadcasting standards is debated, the NHK system is making inroads as a substitute for film in studio production. The use of HDTV videotape in place of film for shooting and editing can reduce some production costs. After being shot and edited on HDTV tape, a program can be transferred to film for exhibition or conventional TV transmission. An industry–government technical standards committee recommended in 1988 that the NHK system become the U.S. standard for future HDTV studio production. The committee's vote was divided, reflecting broadcasters' and manufacturers' concerns about adopting a standard incompatible with the existing broadcasting system. However, it also illustrates that HDTV is being used increasingly for production applications.

In Japan HDTV production began with demonstration tapes. It now includes NHK production of dramas, documentaries, and concerts and is looking toward the mass production of receivers and HDTV satellite-to-home broadcasting in three or four years.

Elsewhere, a few production companies, including some independents in the United States, are using NHK equipment to produce HDTV music videos, commercials, and other short-form projects, which have been transferred to film for broadcast. Some are making a permanent commitment to the new technology. For example, Osmond Studios announced a complete conversion to HDTV production in 1988 with plans to transfer the final product to 35mm film.[6]

With an eye toward reducing production costs, some theatrical film production studios have experimented with HDTV, and numerous demonstration tapes have been made. A fourteen and a half–hour Canadian TV miniseries and a big-budget English-language Italian movie, produced using the NHK high-definition video system in 1987, were among the first large-scale efforts.

The major American studios, while attracted by the economies of HDTV production, still hesitate to commit to it. Although HDTV approaches 35mm film performance in many respects, it is a less than perfect substitute, especially in the eyes of many traditional film artists and technicians. HDTV field production requires bulkier equipment than the usual 35mm or 70mm film cameras plus the presence of a television production van. In the eyes of cinematographers HDTV does not have the same look or feel aesthetically as film. Furthermore, though HDTV compares favorably to film when both are seen on a TV receiver, it must be projected to be seen on a cinema-size screen. Projected HDTV does not yet offer the visual impact (brightness or contrast) that movie goers have become accustomed to. Such considerations and union concerns about lost jobs and retraining may override economic advantages in theatrical film production for some time to come.[7]

Alternative HDTV Distribution Systems

An NHK official has estimated that 100,000 HDTV home receivers may be on the market by 1990 and 500,000 by 1991.[8] Japan's commitment to receiver production suggests that HDTV may first find its way into some homes by routes other than terrestrial broadcasting: satellite-to-home broadcasting, home video, and cable.

Because it uses different frequencies than terrestrial television broadcasting, satellite TV may provide one of the first distribution channels for HDTV. In 1982 CBS, an early promoter of HDTV, received FCC approval for a plan to build a satellite for HDTV transmission. It proposed three different HDTV channels: one for high-definition programming for CBS affiliates, a pay channel for HDTV movies, and a third for special distribution services, including transmission of HDTV films to electronic cinemas. CBS later abandoned the plan, but it demonstrated the feasibility of satellite-transmitted HDTV. More recently, Japan has announced plans for two HDTV broadcast channels to begin transmission in 1990 and a third in 1991.[9] Europe's Eureka system is designed with satellite distribution in mind.

Assuming HDTV VCRs and videodisc players could be marketed at prices comparable to present videophile equipment, home video could provide another medium for HDTV expansion. If adequate software were available for sale or through rental stores, an HDTV home video industry could exist entirely separate from HDTV broadcasting. It would not be necessary to produce new software; the opportunity to see current and vintage Hollywood films in stereo high definition might well provide adequate incentive for equipment purchase.

Trials have already demonstrated the technical feasibility of transmitting HDTV signals to subscribers over conventional cable systems. Using NHK's MUSE transmission system, an HDTV channel requires no more bandwidth on a cable system than the equivalent of two conventional cable channels. The appeal of pay-per-view programming on cable could be dramatically enhanced by high-definition films and live productions such as concerts, Superbowls, of championship boxing. The cable industry's interest in HDTV is growing. The National Cable Television Association has established a committee on HDTV, and Home Box Office, the leading pay-movie service, has taken a role in exploring HDTV cable applications.

Cable interest in HDTV distribution may also be driven by the possibility that HDTV may find its way into homes over telephone lines. The fiber optic lines and digital transmission systems being installed by telephone companies are capable of transmitting video as well as voice and data. A small number of prototype systems already deliver NTSC video to cable subscribers in the United States. If regulators allow it, HDTV could be offered by phone companies to subscribers in the future, posing a significant challenge to the cable industry.

Notes

1. An excellent overview of the development of TV beginning in the 1920s is John P. Freeman, "The Evolution of High-Definition Television," *SMPTE Journal* (May 1984): 492–501.
2. "HDTV Wows 'em in Washington, Too: FCC Majority Gives It a Rave," *Broadcasting* (1 March 1983): 36.
3. "NBC's One Channel Solution to HDTV," *Broadcasting* (10 October 1987): 35.
4. See "Eureka HDTV System Unveiled at IBC Conference," *Broadcasting* (3 October 1988): 66.
5. For an overview of European HDTV plans, see Simon Baker, "Nothing But Trouble?," *Cable & Satellite Europe* (June 1988): 28.
6. "Jimmy Osmond's OCG to Install HDTV Gear at UT Studio Complex," *Multichannel News* (11 January 1988): 30.
7. See "Hollywood on HDTV: Enthusiastic But Cautious," *Broadcasting* (26 October 1987): 76.
8. "NHK Engineering Official Challenges HDTV Critics," *Multichannel News* (21 December 1987): 15.
9. Minoru Inaba, "Japan Begins Promotion Push for High-Definition TV," *Multichannel News* (14 September 1987): 15.

9

Low-Power Television

In 1982 the Federal Communications Commission took steps to expand U.S. television in an unprecedented way. It established the first new broadcasting service in twenty years, an entirely new class of television stations: low-power TV (LPTV). LPTV first had been proposed in the latter days of the Carter administration under FCC chairman Charles Ferris. The objective was to provide a regulatory framework promoting an alternative television industry, one owned by people and groups for whom full power–TV station ownership was beyond reach: ethnic and racial minorities, rural communities, nonprofit organizations, even small-time entrepreneurs. Today LPTV is still in its pioneering stage in the United States, although some promising beginnings have been made. Low-power television is being used by government in countries outside the United States to bring entertainment and education to remote areas. In Italy a form of commercial LPTV is challenging the traditional state-run full-power television system.

Early Low-Power Broadcasting

During the first decade of television station licensing and construction in the United States, many operators of full-power stations found they had troublesome pockets of poor reception in their viewing areas. These coverage problems were often due to terrain features, as when a mountain blocked reception. Other operators wished to extend their signal coverage to communities just outside their primary signal coverage area. The solution to both problems was to build a second booster transmitter near the fringe audience. The transmitter was equipped with a receiver to pick up the station's primary signal, which was then reamplified and broadcast at low power to the nearby audience. Many of these boosters shifted or translated the primary signal to another channel before rebroadcasting it; they became known as translators.

Around a thousand boosters and translators were operating across the country without licenses when the FCC began taking steps to regulate them in 1956. By 1960 the Commission had set up a licensing scheme for VHF and UHF low-power translators. These rules remained essentially unchanged for the next twenty years. They prohibited translators from acting like a full-power station in any way. Other than certain kinds of emergency announcements and a few other exceptions, translators could not originate any programming; they were strictly limited to functioning as a relay service for full-power stations.

LPTV Service Begins

By the late 1970s over 4,000 translators were operating across the United States. The availability of satellite program distribution, more affordable video production equipment, and more powerful but less expensive signal transmission technology made program origination economically possible for translators. The FCC had authorized local origination experiments in several cities and, encouraged by the results, went on to establish LPTV as a separate new television service in 1982.

The FCC was overwhelmed by thousands of LPTV license applications. They included small-town radio station operators, major corporations, urban neighborhood associations, school systems, pay-TV entrepreneurs, religious groups, and other nonprofit organizations. A computerized processing system was installed in 1983 and the FCC stopped accepting further applications. A lottery was set up to select the winners in cases where multiple applicants vied for the same frequency. In the lottery, minority applicants and those with few or no other media properties are given a statistical advantage in accordance with the FCC's goal of using LPTV to increase the diversity in American broadcasting. Although the FCC received some 40,000 applications in all, by 1987 the backlog was finally under control and it began accepting applications again—this time requiring a $375 fee with each one.

In 1988 more than 500 LPTV stations were on the air, 240 of these making up the Alaskan educational television network. In the continental United States 318 stations were broadcasting, and construction permits had been granted for more than a thousand others. Forty-two states had at least one station licensed; nine states had ten or more. The most stations were in California (22), Texas (16), and Wyoming (16).[1] About 60 percent of LPTV stations used VHF channels (2 through 13), the remainder UHF (14 and higher). By contrast about 5,000 translators and 1,000 full-power stations were in operation.

LPTV Rules

The LPTV rules allow translator operators to begin originating programming after simply notifying the FCC. The rules also allow persons who want to build LPTV stations from scratch to do so using translator technology and local origination equipment. LPTV licensees are subject to few of the restrictions the FCC places on full-power stations. Ownership requirements are liberal, financial qualifications are minimal, and there is no set minimum number of on-air hours or even a requirement that a studio be constructed. There is also no formal limit on the number of LPTV stations one person can own, or prohibition against cable ownership of LPTV stations.

However, offsetting the regulatory freedom enjoyed by LPTV is its status as a secondary broadcasting service. As a secondary service LPTV must concede to full-power stations in questions of frequency allocation or signal interference. For example, if an LPTV station's signal interferes with audience reception of a full-power station's signal in a given market, the LPTV station will be required to make technical changes to accommodate the full-service station. Likewise, an LPTV station can be required to surrender its frequency to a new full-power station if that frequency is the only one available in the market. LPTV broadcasts also may not interfere with cable system transmissions, and there is no requirement that cable systems carry local LPTV stations.

Station Costs and Signal Coverage

The character of LPTV operations is set by the transmitter power limitations. Low power means small audience, and small audience means small business. Running an LPTV station that originates programming and sells advertising is very much like running a small radio broadcasting station. Making a profit—or at least a living—with such limited audience reach means constantly finding ways to economize; it's television broadcasting on a shoestring budget.

As a rule of thumb it costs about $250,000 to equip an LPTV broadcasting facility with a transmitter and basic studio equipment. About one-quarter to one-third of this amount is spent on the transmitter, the single most expensive piece of equipment. Add the cost of land, a building, and perhaps a mobile production van, and the average price tag for an LPTV station equipped for local studio production reaches $500,000. Minimum start-up cost for a full-power station is easily five times this amount.

As with any transmitter, the signal coverage provided by an LPTV transmitter depends on its power (rated in watts) and its height above the surrounding terrain. VHF stations are normally limited by the FCC to 10 watts of power, UHF to 1,000 watts. As Table 9–1 shows, this power differential gives them roughly comparable range. Most LPTV stations operating today have signal coverage of five to twenty miles, although in densely populated areas less coverage can be adequate, and some rural stations in flat terrain reach somewhat farther. By contrast, full-power station transmitters typically range from 50,000 to 100,000 watts, with some reaching as far as eighty miles. The transmitters alone often cost more than an entire LPTV station.

Patterns of LPTV Operation

According to the LPTV industry's primary trade group, the Community Broadcasters Association,[2] several distinct strategies or approaches to

Table 9–1. Maximum LPTV Signal Coverage*

	Transmitter Power	*Coverage Distance*
VHF (Channels 2–13)	1 watt	11 miles
	10 watts	19.5 miles
UHF (Channels 14–83)	10 watts	9 miles
	100 watts	18 miles
	1000 watts	26.5 miles

*Maximum distance in miles that a low-power TV station signal will provide a useful signal, using a 1000 foot-high transmitting antenna.

LPTV operation are represented among the several hundred stations on the air:

Educational Network. About 240 LPTV stations comprise the University of Alaska's statewide educational network. The network was begun in 1973 with an FCC waiver of the translator rules that were then in effect. Programmed in cooperation with the University of Alaska, the network reaches remote villages all over the state with educational materials. Many stations also originate programming.

Satellite Downlinks. Many LPTV licenses have been granted for transmitters that rebroadcast to a local audience signals picked up by a satellite dish adjacent to the antenna. Because there is no cost for a studio for local program origination, this approach provides an economical way to bring pay-TV or other entertainment services to homes beyond the reach of cable companies. Nearly half the 300 LPTV stations in the continental United States operate solely as satellite downlinks, according to the CBA. Many of them are licensed to Trinity Broadcasting Network, a California-based religious programming organization.

Small-Town TV. Perhaps one-third of the LPTV stations are engaged in various forms of small-town TV pioneering, in many ways similar to small-town radio—with pictures added. The first of these pioneer stations was channel 26 in Bemidji, Minnesota, a town of 12,000 that is 150 miles west of Duluth. From 1981 to 1983 channel 26 made a go of it with local origination consisting of studio programs, news, and syndicated programming, then switched to local news breaks inserted into a scrambled pay movie service. Other examples typical of this approach include channel 54 in Bucyrus, Ohio, which began in 1984 and turned a profit about two years later. The station offers a mix of local news and sports, movies, and two satellite networks: Tempo and the Cable News Network. Channel 10 in Woodstock, Virginia, offers two daily local news broadcasts, talk shows, and coverage of local events such as parades and even birthday parties. In Tallahassee, Florida, a full-power station owner operates a low-power station in the same community to offer more in-depth coverage of local news and events. Because much of the full-power station's audience is in Georgia, its news and public affairs coverage is necessarily divided between news in the two states. The LPTV station is used to supplement the other station's coverage.

In many small towns, local LPTV programming has been met with enthusiasm and pride in "our own TV station." Although station staffs are small and the technical quality can be less than slick, many LPTV operators find ready advertising support from local businesses.

Urban LPTV. The majority of LPTV stations are in smaller communities, but by 1988 most of the twenty or so largest cities in the United States had stations in operation or under construction. While rural stations base their appeal on providing their communities their first local television service, urban LPTV operators face an entirely different situation. With several television stations and cable service already in place, the urban operator must identify some audience segment or programming need not being served by the major local media.

This audience specialization approach can take the form of non–English language programming or programs designed to meet the information needs of minority ethnic groups. LPTV Spanish-language programs have been successful in Washington, D.C., and other cities, for example. Specialized programming also includes types of information related to certain occupations or businesses. For example channel 55 in Milwaukee programs financial information to the business community during the day and black-oriented and Spanish-language fare at night. Another low-power station in the same city, channel 8, is attempting to draw a younger audience with a rock video format.

Programming Problems and Possibilities

The availability of programming is a critical problem for the LPTV industry, especially for stations in small towns dependent on advertising for revenue. Local production of all programming is beyond the financial capabilities of most stations. The approach taken by most is to produce key programs such as news or local sports coverage, then find affordable programs to fill in the rest of the day. Satellite network services are being used in some cases, as is syndicated and barter programming. The program service used by a station depends on its cost and the number of thirty- and sixty-second slots the program provider leaves for the station to insert its own commercials. Music videos are a popular fill-in on many LPTV stations because they are available free and their short length provides many potential spots to place local ads—much the same as popular music singles do on radio.

As more LPTV stations go on the air and the industry becomes a larger market for program providers, various program packages or services are appearing that are tailored to the needs of low-power broadcasters. For example, the Community Broadcasters Association announced the formation of a program cooperative in 1987 to enable operators to purchase programs as a group, helping to reduce the cost.[3] A national

LPTV programming network, Channel America, has also been proposed, which would provide a three-hour daily block of LPTV programming including news and game shows.[4]

Closely related to the problem of program acquisition is the question of an LPTV station's relationship to cable systems in its viewing area. Cable carriage effectively equalizes the difference in signal strength between low-power and full-power stations, expanding the LPTV audience. Although cable systems are not required to carry low-power outlets in their franchise area, many do. The CBA estimates that of the stations in small communities that have cable systems about 70 percent are being carried.[5] Generally, however, LPTV has not yet made much of an impact on the cable industry. Many LPTV operators are paying a monthly fee to be carried on local cable systems, others have been refused carriage outright, especially in urban areas.

LPTV: Surviving a Traumatic Birth

When asked to describe the state of the industry in the late 1980s a consultant remarked, "I think LPTV has survived its traumatic birth."[6] The trauma referred to was the nearly four-year freeze on new LPTV licenses from 1983 to 1987 needed to deal with the flood of applications. Many industry observers feel that period of uncertainty and frustration waiting for licenses took some of the momentum out of LPTV. Many proposals were set aside as enthusiasm faded and financial backers found other ventures.

As is the case with any newborn, it is difficult to say what the future holds for LPTV. Clearly it has the potential to provide local television programming for many types of audiences that have been by-passed by commercial TV and, to a large extent, by cable also. As is nearly always the case with new media in the United States, the question is ultimately one of economics. Small-town LPTV operations, the pioneers of the industry, have been enthusiastically received in many communities, and some are actually making money. Many more have not yet found the right combination of local production and affordable nonlocal material to make a go of it. Like so many small business enterprises, many fledgling LPTV stations will succumb to undercapitalization, running out of money before they start making it.

Urban LPTV stations, many serving linguistic and ethnic minority audiences, may face a somewhat different challenge. As the value of television stations in urban areas continues to rise, the LPTV channels in these areas will become even more attractive to large media companies. The possibility exists that these specialized stations carrying local programming might be sold to firms developing more commercialized stations, national networks, or simply to speculators. Either way, they would lose much of the "peoples' TV" character envisioned by the FCC when it established LPTV.

Notes

1. "Facts and Figures: Low Power TV Distribution," *Multichannel News* (11 May 1987): 45.
2. Jacquelyn Biel, Community Broadcasters Association, interview 30 December 1987.
3. Janet Stilson, "Low Power Group Seeks Programs," *Electronic Media* (17 August 1987): 26.
4. Janet Stilson, "Low Power Network to Premier," *Electronic Media* (17 July 1987): 24.
5. Andrew Grossman, "LPTV Broadcasters Increase Efforts to Be Carried on Cable," *Multichannel News* (17 August 1987): 57.
6. Biel, interview.

10

Subscription Broadcasting: STV and MMDS

Since the beginning of commercial television broadcasting, various companies have experimented with ways to bring pay TV into subscribers' homes without the expensive and time-consuming construction of a wired distribution system. Today three forms of over-the-air pay television have evolved from these efforts: subscription TV (STV), multichannel multipoint distribution service (MMDS), and satellite-to-home broadcasting. MMDS and STV are terrestrial media that use short-range, earth-bound television broadcasting technologies for local program distribution. Satellite-to-home video, the newest of the three, is a national distribution medium. It is dealt with in Chapter 5.

STV: The First Pay-TV Medium

Subscription television (STV) refers to any system used to scramble a conventional television station signal so it can be received only by paying subscribers using a descrambling device. STV dates back to the beginnings of the television industry in the late 1940s when Zenith Radio Corporation began experiments with a pay-TV service called Phonevision. It was called Phonevision because the signal used to unscramble the television picture reached subscribers via phone lines.

During the 1950s and early 1960s the FCC authorized several STV experiments including one that used coin boxes attached to TV receivers and another in which punch-card tickets were used to activate the decoder. The biggest Phonevision experiment involved WHCT-TV, a Hartford, Connecticut, UHF station, which offered movies and other programming from 1962 to 1965.

The Hartford experiment led to the FCC's first rules authorizing regular STV service in 1968.[1] After twenty years of experimentation and regulatory delay, the STV industry finally had arrived. However, in an effort to protect the economic interests of local television broadcasting, especially the struggling UHF stations, the FCC made the STV rules quite restrictive. STV stations had to offer a minimum of twenty-eight hours of unscrambled free programs per week. They could not offer movies more than three years old or sports events that had been broadcast on free TV within the last five years. Operators were required to lease decoders to their subscribers, not sell them. Finally, an STV station could operate only in a community where four conventional stations already existed. The rules effectively confined STV to only the country's larger markets and prevented it from competing with existing stations for programming; by 1977 only two STV stations were on the air, both in Los Angeles.

STV's Sudden Rise and Fall

That year the industry received help from an unexpected source. HBO, the cable industry pay-movie service, won its court fight to have removed similar program restrictions that had been imposed on the cable industry.[2] Consequently, the FCC also deleted STV's program restrictions. STV operations began to appear in major cities around the country and, as industry growth accelerated, a new deregulation-minded FCC removed other restrictions as well. By 1983 all the old STV rules had been set aside.

In 1982 twenty-two major cities had STV operating on UHF stations. Some, like Los Angeles or Dallas–Fort Worth, had two or more services. The Los Angeles–based services, ONTV and SelecTV, provided sports and movie packages via satellite to operators across the nation. Total national subscribers climbed to 1.5 million. Then over a few months the STV industry collapsed dramatically. In 1983 subscriber growth turned into loss as stations dropped their STV operations. By 1985 the industry was only half its 1982 size; by 1988 it had dwindled to the original Los Angeles operations.

Reasons for the Fall

A number of problems contributed to STV's collapse as a major player in the pay-TV industry, including economic recession and signal piracy. Also a factor was the increasing profitability of UHF stations, which made advertiser-supported broadcasting more attractive. It is clear, however, that the major culprit was the slow but sure penetration of cable into major urban areas in the early 1980s. Although there were a few loyal exceptions initially, once cable reached STV subscribers, it was difficult for them to resist thirty-plus channels for about the same price as STV's single channel.

The STV story might have been quite different. Had the FCC's 1968 STV rules been more liberal the industry might well have become established in major cities by the early 1970s when cable was still mostly a suburban/rural medium. Then, when satellite movie programming first became available in the mid-1970s, STV would have been in position before cable to become the nation's urban distributor of these services. As Chapters 11 and 12 indicate, without the revenue potential of satellite pay programming, the cable industry probably would not have gotten the financial backing necessary to build the major urban systems that led to STV's downfall. However, as STV's role as a major player on the pay-TV scene was coming to an end, a new pay broadcasting technology, MMDS, was gearing up to challenge urban cable.

MDS: Predecessor to Wireless Cable

In 1962, while still mulling over the future of STV, the FCC approved a new kind of television broadcasting service that would eventually take STV's place in the urban pay-TV business. Ironically, Multipoint Distribution Service (MDS) did not appear to have any potential as a mass entertainment medium at the time it was approved.

MDS was one of three services the Commission designated to share a block of high-frequency microwave channels in the vicinity of 2 gHz. The other two were operational fixed service (OFS) and instructional television fixed service (ITFS). OFS frequencies are licensed to businesses for relaying voice and data communications from one site to another within a company. Only internal information owned by the company can be transmitted. ITFS frequencies are set aside for the transmission of educational and cultural television programming by school systems, universities, and similar organizations. Lectures, demonstrations, training courses, and adult educational programs typify ITFS fare.

MDS Transmission Systems

Although most microwave communication systems use narrow beams of energy to relay information from point to point, an MDS system broadcasts an omnidirectional signal that can be picked up by any number of receivers. In this way MDS transmission resembles conventional television broadcasting; it just uses much higher frequencies. Conventional TV channels are all located below 1 gHz (between 76 and 806 mHz). Because higher frequencies are used, a special antenna is needed for reception; it is smaller but more expensive than the typical TV rooftop antenna.

After reception, the high-frequency signal must be shifted to a lower-frequency channel that a home TV receiver can tune in. This is done by a downconverter, a small black box mounted on the antenna mast or inside near the TV set. The MDS signal is downconverted to a regular TV channel not being used in the community.

The usual MDS signal carries about twenty-five miles. Unlike conventional TV, an MDS signal is limited to line-of-sight reception: the receiving antenna must be able to "see" the transmitter—no objects may block the path. Extremely hilly terrain or tall buildings can pose problems for MDS reception.

MDS Growth and Transition

The MDS system approved by the FCC in the 1960s attracted very little interest as a potential entertainment medium. The first channels set aside were not adequate for color transmission. Also, antennas and downconverters were too expensive at that time for home service to be practical. The FCC modified the MDS rules to permit color transmission in 1970 and then established two MDS channels in most major cities in 1974.

In 1975 HBO became available by satellite, and advances in microelectronics reduced the cost of MDS transmission and receiving equipment. MDS stations sprang up rapidly in major cities across the country offering HBO and other services to subscribers not yet reached by cable. Customers also included motels, hospitals, hotels, and apartment complexes. By 1982 more than 300 MDS stations had been licensed by the FCC, and MDS pay-TV systems were operating in over seventy cities, serving some 750,000 subscribers.

The Establishment of Wireless Cable

Like STV, MDS was looking over its shoulder at the growing urban cable industry. As cable systems were completed in major cities, MDS subscribers switched to cable, and the national subscription level began to decline. MDS operators feared that their industry could not survive as a one-channel medium. One of the major operators, Microband, proposed that the FCC allocate more MDS channels, to allow MDS to compete more effectively with cable. In 1983 the FCC, following Microband's suggestion, allocated eight more MDS channels to every major city.[3]

The channels were assigned in two four-channel blocks, taken from the seven four-channel blocks set aside in each city for ITFS. Each major city now had eight channels plus the original two MDS channels available. In addition to the new channels, the FCC changed the ITFS rules to allow MDS operators to lease unused time on ITFS channels, making it possible for an operator to offer a dozen or more program channels to subscribers. This new, improved MDS was called multichannel multipoint distribution service (MMDS). Microband dubbed it wireless cable.

Within a few months more than 16,000 applications were received. As it did with low-power TV, the Commission established a lottery procedure to award MMDS licenses in cases where there were multiple competing applications. However, a two-year delay ensued during which the FCC was embroiled in a debate over awarding lottery preferences to minority applicants and which FCC bureau should have jurisdiction over the new service. During the delay many single-channel MDS operators lost new subscribers who had been signed up with the promise of multichannel service to come. After a year about fifty frustrated MMDS applicants staged a walk-out at a public meeting of the FCC when the chairman announced once again that MMDS had been deleted from the agenda.[4]

It was during this delay that the first urban MMDS system appeared, thanks to the changes in the ITFS rules. Using four channels leased from George Mason University, the Capitol Connection offered pay movies, sports, and news services to Washington, D.C., subscribers for $35 per month.[5] Finally, late in 1985 the first MMDS lottery was held assigning the new four-channel blocks in cities across the nation.

The following year the FCC gave the industry another boost by substantially deregulating it. From the outset MDS had been classified as a common carrier. Any medium classified as a common carrier is prohibited from owning or controlling the content of the transmissions it carries. All telephone services in the United States are regulated as common carriers, for example. The rates charged by common carriers must be approved by the FCC and cannot be changed without permission. In 1986 the FCC gave MDS/MMDS licensees freedom to choose either common carrier or non–common carrier status. This gave MDS/MMDS operators roughly the same flexibility in offering pay-TV programming as cable systems, STV, and DBS.

In 1988 fifteen MMDS operations had been licensed in major cities, although fewer were actually operating. According to the industry's trade group, the Wireless Cable Association, about 250,000 MMDS subscribers were receiving service, most of them in a small number of major markets, including Cleveland, Detroit, Washington, D.C., and New York. About 100,000 more people were served by a few dozen single-channel MDS systems still operating, most of them in midsize to smaller markets.[6]

MMDS Operations

After receiving a license from the FCC, a prospective MMDS operator will plan to spend from $100,000 to $250,000 to construct a transmission facility. This investment is comparable to LPTV start-up expense and only a fraction of what a full-power TV station or cable system typically costs. Its variability depends on a number of factors, including whether a tower must be built or can be leased, the number of channels involved, and the power of the transmitter.

However, the transmission facility cost is only a small part of what must be invested to establish a wireless cable pay-movie operation in a major market. Operators in major cities have invested as much as $10 to $15 million to mount a full-scale wireless cable operation. In addition to decoders and receiving antennas (and the crews and equipment for their installation), there is the cost of programming, fees for ITFS channel use, administrative and marketing staff, and the considerable expense of an advertising and promotion campaign in a major city.

An MMDS licensee can either assume these expenses directly or make an arrangement with a separate company to finance and provide programming. Another alternative is to operate as a common carrier, with no control or involvement with programming, content to collect FCC-set transmission fees from the pay programmer. This reduces the licensee's risk and capital outlay but eliminates any potential profit from subscription revenues.

The number of channels and the programming packages on MMDS systems can vary considerably from one operation to another. Under FCC rules, no one licensee can control more than one of the two four-channel MMDS blocks in a market, or one of the two single-channel MDS frequencies. Consequently a wireless cable system might use four channels under one licensee's control, or eight controlled by two licensees, or twelve channels, some—or all—of which are leased ITFS facilities. To put together a wireless cable system with a dozen or more channels requires considerable cooperation and negotiation among several parties.

Access to Programming and Competition with Cable

In many cases, lining up the channels is just the first challenge for a wireless cable entrepreneur; finding affordable programming can be a major difficulty. Depending on commitments already made to the local cable system operator, pay-movie channels and other satellite program services can be very reluctant to deal with MMDS operators. For example, in any city where the local cable system is already an HBO client, why should HBO be interested in selling movies to an MMDS operator competing with the cable system? Because the number of subscribers affects the market value of a cable system, loss of current or potential subscribers to MMDS is a serious concern to cable operators. They do not look favorably on the prospect of HBO or any other satellite service providing programming to their competition.

On the other hand, MMDS operators do not see their service necessarily as a threat to cable but as a technology that can reach the people cable cannot. They argue that because of the difficulties of cable system construction and the expense involved, there will always be potential subscribers left unserved by cable, even in major metro areas. They say MMDS and cable should peacefully coexist and cooperate, to the benefit of both. They accuse the pay-movie services and cable industry of collusion against MMDS and in early 1988 were seeking federal legislation to prevent programmers who sell to cable from discriminating against MMDS in the same markets.

Programming Security Concerns

Whether single or multichannel, MDS operators must deal with the threat of signal piracy, the unauthorized reception of their pay movies and other premium programming. In the late 1970s when single-channel MDS was rapidly expanding around the country, many operators did not use any form of scrambling to protect their signals. The need to locate and purchase a special antenna and downconverter seemed to be adequate deterrent to most potential eavesdroppers. However, once people discovered that MDS receiving antennas and downconverters were readily available through local electronics supply houses and catalog companies, unauthorized viewing became a major problem. For example, in Los Angeles pirates were said to outnumber subscribers by seven to one.[7]

Although the law stipulates that unauthorized reception of unscrambled pay TV is illegal,[8] the use of security systems by MDS/MMDS operators has become common. The basic approach is to scramble the signal in much the same way as an STV signal is scrambled. In addition to the downconverter the subscriber must also be provided a descrambler to make the picture viewable. For more security addressable descramblers can be used that must receive a special code in the picture signal before they will descramble it. Each descrambler has a unique code to respond to, and that code is not transmitted by the station unless reception is paid for. Similar systems are in use by satellite-to-home broadcasting services.

Wireless Cable Outside the United States

At present MMDS is almost exclusively a U.S. phenomenon, although there is considerable interest in it elsewhere. According to a major MMDS equipment manufacturer,[9] an MMDS system in the Bermuda islands was one of the first established in the Western Hemisphere. Curaçao (an island off the coast of Venezuela) can claim the distinction of having the only other operational system outside the United States. Both systems serve subscribers and hotels and other establishments associated with tourism. In Australia a single-channel MDS system transmits movies and other entertainment fare to commercial establishments but is not yet licensed to serve home subscribers.

For a country to have MMDS, it must first have the proper political and economic environment for such a medium. Obviously, in many countries a four-channel or even single-channel microwave entertainment or information service is either irrelevant or economically impractical. However, in areas where there is either an economic or political rationale for multichannel television, MMDS is seen as an attractive alternative to the expense of cable system construction.

Countries interested in developing their own MMDS systems then must manufacture their own transmission equipment or purchase it from the United States, the only place where it is available commercially. U.S. MMDS transmission equipment, however, will operate only on U.S. MMDS frequencies. In many countries these frequencies are already in use by other microwave media. For example, cable operators in Great Britain have requested permission from the Cable Authority to establish MMDS facilities to serve customers while major cable systems are under construction. Because the needed frequencies are already in use, it is not likely they will be made available for British MMDS.

Consequently, in Britain and elsewhere, for U.S.-type MMDS to develop, problems of frequency availability and transmission equipment standards will have to be solved. The only alternative is for each country desiring wireless cable to find a new band of unused microwave frequencies, designate them for this service, and find a manufacturer to produce a transmission system for those channels.

Notes

1. "Fourth Report," 15 F.C.C.2d 466 (1968).
2. HBO v. FCC, 567 F.2d 9 (1977), *cert. denied,* 434 U.S. 829 (1977).
3. "Multichannel MDS Is Here," *Broadcasting* (30 May 1983): 32.
4. Stephen J. Shaw, "MDS Applicants Walk Out to Protest FCC Delays," *Private Cable* (December 1984): 30.
5. J.L. Freeman, "First Urban MMDS Op Launches," *Multichannel News* (17 December 1984): 1.
6. Curt Bradley, president, Wireless Cable Association, interview, 20 January 1988.
7. "Chasing MDS Pirates," *Broadcasting* (1 November 1982): 59.
8. Title VI, Communications Act of 1934, as amended by the Cable Communication Policy Act of 1984, Public Law 98-549, 30 October 1984.
9. Jim Clark, Conifer Corporation, interview, 20 January 1988.

PART IV

The New World of Wireline Media

11
Cable TV Growth and Regulation

Soon after the first television broadcasting stations went on the air in major U.S. cities in the late 1940s, cable systems appeared in towns just beyond the reach of those stations. Cable as we know it today has evolved from that simple extension of local TV to a significant competitor for network television audience in the United States and an increasingly important medium in Western Europe and elsewhere. This chapter provides highlights of that evolution and an outline of its regulatory context. Chapter 12 explains how local cable systems are built, operated, and programmed.

First Phase: An Extension of Local Television

In the late 1940s commercial television broadcasting stations were rapidly going on the air in big cities across the United States. The demand for licenses was such that it exceeded the FCC's TV frequency allocation plan. While it worked on a new plan for additional TV channels, the FCC imposed a licensing freeze. During the freeze, from 1948 to 1953, the number of pioneering television stations stood at 107, and all were located in larger cities.

During the freeze the U.S. television industry was being born. Increasing numbers of TV receivers began to appear in stores and households within range of existing transmitters, as local stations and their networks experimented with and developed programming. The excitement and curiosity surrounding TV in the larger markets spread rapidly to outlying areas where would-be viewers and local appliance dealers were soon frustrated by poor fringe reception or no reception at all. In many communities enterprising set owners, business people, and even local governments cooperated in erecting various kinds of master antennas on towers or on nearby hills to make TV reception possible. Among the first such community antenna systems was that built by an appliance dealer in Lansford, Pennsylvania, in 1949. By erecting an antenna on a hill outside town, Robert J. Tarlton was able to pick up three stations from Philadelphia, about seventy miles away. The signals were amplified and passed by coaxial cable to subscribers in town. Tarlton's company, Panther Valley Television, is often said to be the nation's first cable operation.

In 1950 a national count of such community antenna TV (CATV) systems revealed fourteen operating; in 1952 there were seventy, most in smaller communities like Lansford just beyond the reach of the new TV stations. After the FCC licensing freeze was lifted in 1953, commercial television grew rapidly and CATV expanded along with it. Although an increasing number of towns had at least one station, given the economic and technical limitations of the developing TV industry, many homes outside the major markets could not receive an adequate signal. Consequently CATV continued to grow unimpeded during the 1950s as a natural extension of commercial broadcasting. Broadcasters generally did not object to having their signal coverage area extended. The FCC, anxious to get the TV industry off the ground, viewed CATV favorably, finding no reason to extend its regulatory authority over what was seen as a useful adjunct to TV broadcasting.

Phase Two: The FCC Takes Control

The Imported Signal Controversy. However, the FCC's and broadcasters' attitudes toward cable began to change in the early 1960s. Cable was finding its way into larger communities with existing TV stations. In addition to offering improved reception of local stations, a number of systems in larger communities imported distant TV signals to bring in the second and third networks or to add independent station programming to their line-up. In some cases the imported signal duplicated local network affiliate programs. In all cases, broadcasters felt that if the growing cable industry were allowed to continue importing programming, local

TV stations would be threatened economically by the diversion of viewers to imported channels.

The First FCC Rules. In the early 1960s cable reached only about 2 percent of U.S. TV households, but the FCC was persuaded by the broadcasting industry's pleas for protection from possible economic damage. The Commission was particularly concerned about the impact of imported signals on struggling UHF commercial and educational stations, the most vulnerable stations in most markets. In 1962 the Commission asserted regulatory control over cable systems importing distant signals. Then in 1966 the FCC extended its authority over all cable systems, whether importing signals or not.[1] The 1966 rules required that all cable systems carry all local stations and that any duplicate imported programming had to be delayed at least one day. The FCC's assertion of control over cable did not go uncontested; one of the country's largest cable systems, in San Diego, California, fought the rules in federal court only to have the U.S. Supreme Court rule in favor of the FCC in 1968.[2] The Southwestern Cable case today remains the foundation of FCC authority over cable.

The 1972 Rules. Throughout the 1960s cable continued to grow, and the percentage of TV households with cable quadrupled to about 8 percent by 1970. Even then, with only about 4.5 million subscribers spread out over nearly 2,500 systems, cable was not considered a major player among electronic media because of its lack of advertising revenue and its tendency to be mostly a smaller-market phenomenon. In the major markets, cable had made little headway, primarily due to the existence of multiple over-the-air signals and generally adequate reception.

From the industry's standpoint, the outlook for growth was further dampened in 1972 when the FCC released its first comprehensive set of cable regulations. The 1972 rules, intended as a definitive statement of the regulatory status of cable, made it clear that the Commission's top priority was to protect television broadcasting from any economic damage cable might do through programming competition. The Rules effectively prohibited cable systems from competing with TV stations and networks in movie and sports programming, required them to provide public, educational, and government access channels, limited the number of imported signals and specified minimum number of channels according to market size, and established federal franchising and technical standards to ensure that restrictive FCC policies were carried out locally.[3]

The early 1970s were a difficult time for the cable industry. Although growth continued at a respectable rate—about 14 percent of the nation's homes had cable in 1975—it appeared that the steady march of cable from rural areas to the big cities would stall in the suburbs. Urban systems would simply cost too much to build, given the revenues the typical cable system could produce. The wired nation foreseen by some seemed unlikely to become a reality.

Third Phase: Satellite Programming, New Growth, and Deregulation

In September 1975 Home Box Office, a little-known cable program distributor, began offering cable systems commercial-free movies transmitted in a unique new way: via the Satcom I satellite launched only months earlier. Although movies and sports events were being distributed to some systems on tape or live by terrestrial microwave, it was an expensive and awkward arrangement to administer. HBO's satellite system allowed potentially every cable operator in the country to receive the service without adding to distribution cost.

Taking off slowly at first, in large part due to the high cost of satellite receiving dishes (as much as $100,000), HBO soon became a cable industry phenomenon. Cable now had a product not available on local TV, commercial-free movies; a rapid, inexpensive way to distribute it, satellite; and consequently a new source of revenue in addition to regular subscriptions, pay-TV fees. Other pay-movie services, as well as advertiser-supported channels and superstations, soon followed. With pay channels bringing in new revenues and advertiser-supported channels stimulating further audience interest and building the overall subscriber base, the cable industry took on a much more promising look financially. Major media companies either expanded their cable holdings or bought in for the first time.

With new revenues and new financial backing the cable industry took up the challenge of wiring the major urban markets so far neglected. With the financial resources in hand to pay the steep price of big-city cable construction, major cable companies began to bid fiercely for urban franchises. In some cases the high-stakes bidding led to unethical, even illegal, attempts to influence or gain favor with local politicians or special-interest groups. Cities responded by bargaining for the most elaborate and advanced systems possible, with as many bonuses included as could be negotiated, including donations to special city projects not directly related to cable service. However, escalating costs of urban cable construction, high interest rates, and disappointing inner-city penetration rates soon led to the sale and renegotiation of many of the most elaborate proposed systems. Consequently, most of the systems eventually built in urban franchises ended up resembling their more modest suburban counterparts in channel capacity and technical sophistication.

Between 1975 and 1985 the cable industry's dramatic expansion was in part made possible by the FCC's gradual loosening of the regulatory constraints it had imposed in 1972 and before. Faced with certain changes in copyright law and with decisions in which federal courts attacked the FCC's rationale for regulating pay programming,[4] access channels, and distant signals,[5] the Commission thoroughly researched the economic relationship between the cable and broadcasting industries in the late 1970s. It concluded that the economic damage from which it had been protecting the broadcasting industry was no longer an issue; local broadcasting was itself a booming industry. By 1980 the FCC had done away with many of the restrictions in the 1972 Rules, including those on distant signal importation and program duplication, and most of the federal standards for franchising and system construction.

In the same vein, Congress passed the 1984 Cable Communications Policy Act, which made many of the FCC's rule changes law of the land. The 1984 Act also reduced the power of local franchising authorities—the cities—to regulate cable system subscription rates and programming or to require access channels. In 1985 the FCC's "must carry" rules requiring cable systems to carry all local stations were struck down by the courts.[6] At this point, with most restrictive FCC rules mostly set aside and the favorable 1984 Act in place, the cable industry entered a new regulatory era.

Altogether, the franchising and gradual wiring of the cities, the growth of satellite programming, and the removal of many federal regulatory constraints made the 1975 to 1985 period cable's greatest expansion era.

Fourth Phase: Cash, Consolidation, and Credibility

During this expansion period, as cable systems were completed in major metropolitan areas, the percentage of U.S. TV homes subscribing to cable climbed rapidly. Subscribers doubled by 1981, jumped another 50 percent by 1985, then reached the 50 percent level in 1987. The 50 percent level was celebrated as an historic benchmark: cable had become a common household convenience.

Along with subscriber growth, the programming networks enjoyed increased revenues. After struggling through the 1970s all the major satellite programming services became profitable in the 1980s and were able to invest in more original productions and bid successfully against the broadcast networks for other programming. A programming milestone was passed in 1987 when ESPN, a sports network, outbid ABC for a package of National Football League games.

In the ratings cable programming was taking its place beside the broadcast networks, its audience increasing each year as the networks's combined share continued to erode. By 1988 cable programming represented 19 percent of total TV viewing nationally and 33 percent of viewing in homes with cable.[7]

At the same time, cable's financial picture appeared increasingly rosy. Many systems built in the previous decade had finally paid off construction loans and, like homeowners who finally pay off a thirty-year mortgage, found themselves with a surplus of cash. Consequently in 1987, for the first time, cable systems sold on average for $2,000 per subscriber, the same amount that TV stations were sold for per viewer that year.

The largest multiple system operators (MSOs) bought up systems around the country and industry consolidation increased, with a dozen or so of the largest MSOs becoming the dominant players. In 1988 the five largest MSOs reached 42 percent of the nation's cable subscribers; the top ten reached 54 percent.[8] Cable industry consolidation and profitability attracted Wall Street investment interest, further increasing the credibility and financial power of the industry. One financial analyst declared that cable was "the best business in America" to buy into.[9]

Having negotiated the turbulent rapids of franchising and growth in the 1970s and early 1980s, the industry now finds itself with additional momentum in somewhat smoother waters. Major urban systems in the New York, Chicago, and Los Angeles areas are under construction, joining the smaller-market and suburban systems already in place. As these major markets come on line, national cable penetration growth will begin to level off as the industry enters a mature stage of development characterized by less dramatic, more incremental growth. The 8,000 cable systems in operation have passed about 80 percent of the nation's 88 million TV households. In 1988 about 45 million households subscribed to cable (51.1 percent), and 26 million (29 percent) took one or more pay services. Cable industry revenues from subscriptions and advertising (about $12.5 billion in 1987[10]) were about half that of broadcast television and growing.

From the mountains of Pennsylvania to the homes of half the nation, cable's forty-year journey has been arduous at times. But today the faith of early entrepreneurs has been vindicated. What began as a mere extension of broadcast TV has finally become a major electronic medium in its own right, with the resources to challenge the very system from which it arose.

U.S. Cable Franchising and Regulation: An Overview

Although commercial television broadcasting is licensed federally and experiences little direct regulatory control at the state and local levels, cable systems are subject to significant local as well as federal regulation and sometimes state control. Local cable systems operate on the basis of a franchise granted in most cases by the city or county they serve.

The Local Franchise. A franchise is essentially a contract between a cable operator and a city setting out the nature of service provided and establishing the ground rules under which the system will be built and operated. Franchises are granted for a set period, usually ten to fifteen years. Federal law prohibits the operation of a cable system without a franchise because cable system construction and operation requires the use of public property such as streets and utility poles.

Typically, four steps are involved in the franchising process. First the city passes a cable ordinance, a law authorizing officials to grant a franchise and spelling out how the process is to take place. An effective cable ordinance provides a broad outline of what the city desires in a cable system and establishes franchising procedures that allow adequate time and public input to make decisions in the community's best interests.

In accordance with the cable ordinance, the city then publishes a request for proposals (RFP) to build and operate the system. The RFP outlines the detailed information the city wishes interested applicants to provide in their proposed systems.

The third step is the evaluation of proposals submitted. Often, outside consultants are called in to weigh the merits of competing proposals, especially the technical aspects of the plans for construction and operation. Open hearings and meetings allow the competing companies to present their plans for the new system to the public and city administrators. Finally, the franchise is awarded to the winning applicant and construction begins. Particularly in large cities, where financial stakes are high and organized lobbying and interest-group politics are common, the franchising process can become highly politicized and controversial. The high stakes nature of some major-market franchising has been offset to a degree by dividing a city into several areas and granting a franchise for each. Chicago and Houston, for example, have taken this multiple franchise approach. Once signed by all parties, the franchise becomes a legal contract between the city and the system operator.

After the franchise is awarded, local government's task becomes that of monitoring system construction and the provision of cable service. When the franchise runs out, the city can either negotiate a renewal franchise with the same operator or drop the old operator and issue a new RFP, if it can make a case that the old operator failed substantially to honor its franchise commitments.

Federal and State Regulation. All local franchising takes place under the supervision of the federal government because federal law and FCC regulations set the basic parameters for the relationship between cities and cable operators. Whie the city and the cable operator play out the details of system operation and construction, the federal government has established the basic rules of the game and the dimensions of the "playing field." The Communications Act of 1934[11] authorizes local governments to grant and renew franchises, establishes the procedures they follow, and sets limits on their regulatory powers.

For example, in markets where there are at least three local TV stations, cities cannot regulate the rates charged by the local cable system for program services. The Act also imposes some uniform requirements of its own on all cable systems, such as making lockboxes available to subscribers to prevent unauthorized viewing of adult channels by minors, and the establishment of commercial leased access channels on systems with thirty-six or more channels.

Under the Act, the Federal Communications Commission is charged with overseeing the cable industry. While an FCC operator's license, such as that required of broadcasters, is not needed to run a cable system, various technical certifications and permits are. All cable systems are registered with the FCC, and their technical operation is monitored to prevent microwave relays or cable transmissions from interfering with over-the-air communications using the same frequencies.

The FCC also enforces specific requirements of the Communications Act, like the adult channel lockbox requirement. Thus, although the FCC cannot pull the license of a cable system in the same way it can shut down a radio or TV station, it can prohibit the operation of technically substandard equipment until it meets certification standards and levy fines for technical and operational violations of FCC rules.

The usual state statutes applying to fair trade, businesses practices, hiring, and so on also apply to cable systems. Many states have specific statutes prohibiting the theft of cable services, making the unauthorized viewing of cable programming a state as well as a federal offense. In six states public utility commissions or other agencies, rather than local authorities, handle franchising; in three states municipalities grant franchises only under state supervision.[12]

Thus U.S. cable regulation is often a trilevel affair, with federal, state, and local governments all involved to varying degrees. On a day-to-day basis, the typical cable system operator's primary concern is local regulation, which involves a constant give-and-take between the operator, the subscribers, and the city agency tasked with overseeing the franchise. The interplay of these three parties is shaped to a large degree by federal laws and FCC regulations. These laws and regulations have been liberalized in recent years, giving cable operators unprecedented control over their programming services and rates, while at the same time limiting local government's authority in these areas.

Furthermore, a series of court cases has brought into question the whole concept of local franchising to the extent that it exceeds the minimum control needed to ensure that cable system construction and operation will not interfere with other more essential services or infringe on property rights. Ultimately this issue of city franchising authority versus cable operator first amendment rights may be decided by the U.S. Supreme Court.

Private Cable: SMATV

FCC cable cable regulations and the Communications Act set limits on the definition of a cable system. For example, any system with fifty or fewer subscribers is exempted from cable regulation. Also exempted are systems that serve only subscribers on private property in multiple unit dwellings (apartment complexes, condominiums, motels, and so on) and that do not string wires across public streets or other public property.

Master antenna systems have been around nearly as long as TV itself and consist of a single, centrally located antenna that distributes TV signals by wire to the tenants of an apartment complex or other multiple unit dwelling. MATV systems first appeared in cities where apartment dwellers found their "rabbit ears" inadequate and individual outdoor antennas impractical. Beginning in 1979, when the FCC deregulated satellite receiving dishes, dishes were added to many MATV systems to add satellite programming services. Satellite master antenna TV (SMATV) caught on quickly in many cities where cable system franchising and construction was proceeding slowly.

Apartment complex owners and other entrepreneurs, recognizing the unsatisfied demand for cable programming, rapidly installed SMATV systems, sometimes not long before the local cable system arrived. Cable systems and cities complained that SMATV systems were cream skimming and threatening the economic viability of the cable franchises. In many instances this was no exaggerated claim. For example, in the Bronx, New York, 15,000 apartments were served by the country's largest SMATV system, which offered thirty channels.

During the 1980s the relationship between cable and SMATV was also marked by antitrust suits and other legal actions stemming from the refusal of many pay program services, like HBO and Showtime, to sell programming to SMATV operators, whom they saw as direct competition for their primary customers, the cable systems. Programmers and cable operators countered with efforts to convince local governments to pass access laws forcing the owners of multiple dwelling unit to allow cable on their premises. However, the FCC has continued to prohibit local regulation of SMATV.[13] The result has been that the SMATV industry has been able to hold its ground—to keep its exempt regulatory status and force or persuade most of the pay services to sell it programming. By 1988 there were about 2,100 SMATV systems serving an estimated 850,000 subscribers.[14]

Additionally, the cable industry's remarkable growth and financial success in the late 1980s has led to a change in tactics: "If you can't beat them, buy them." A number of major MSOs have bought out or merged systems with SMATV operators in their markets. Many SMATV operations have become a part of the larger franchised cable operations that surrounded them, to the financial benefit of both.

Cable Outside the United States

Cable has developed to some degree over the last thirty years in most industrialized countries. Outside the industrialized nations cable exists, if at all, in only the largest urban areas. National penetration figures can be misleading because of great differences in the age, channel capacity, programming restrictions, and regulatory status of cable among nations.

For example, in most European countries the term *cable* is used interchangeably with MATV because of the widespread use of master antennas in multiple unit housing.

Most cable systems were developed for the purpose of retransmitting national program services, not importing programming, and often older systems have only a limited capacity, perhaps four channels. Federal policies, not local, traditionally dictate the availability of programming, often prohibiting local or imported programming that would compete with the national services.

Traditionally, cable systems outside the United States are under control of a country's telecommunications ministry or PTT, the government-run post-telephone-telegraph monopoly. Local governments request cable service and the PTT provides or authorizes it. The PTT may install and manage individual systems at the local level as part of its telecommunications monopoly or may issue licenses to local operators, which may be the local government or private companies in partnership with local authorities. However, the growth of pan-European satellite programming and the trend toward television deregulation in Europe and elsewhere are leading to liberalization of cable regulation and programming in many countries.[15] For example, in the United Kingdom the British Cable Authority now accepts applications from private companies for cable franchises in major cities and awards them in a way somewhat similar to the U.S. local model on a national scale.

Reportedly the most heavily cabled country in the world is Belgium, where 88 percent of households have cable. The average Belgian cable viewer can receive sixteen channels. Canada can claim second place with 76 percent of homes wired (80 percent of TV households have been passed). Approaching Canadian penetration levels is the Netherlands, with about 70 percent. However, much of this figure represents old, low-capacity systems built initially in the 1960s to provide national radio relay service. Amsterdam has Europe's largest cable system with 354,000 homes. In Denmark and Switzerland cable reaches about 60 percent of TV households, and about 50 percent in Sweden.

Several of Western Europe's wealthier nations have low cable penetration, in part due to well-developed terrestrial broadcasting systems and availability of spill-over broadcasting from neighboring countries. France has only 5 percent, although a national plan to wire the country with state-of-the-art technology is underway and has a goal of 50 percent coverage by the mid-1990s. In Paris, construction of a two-way fiber optic cable system began in 1986. West German cable penetration is around 8 percent and Great Britain only about 1 percent. Britain, however, is in the process of developing franchises in several urban areas including Birmingham, which when completed would reach nearly 450,000 homes. Other European penetration levels include Ireland, 30 percent; Norway, 25 percent; and Austria, 11 percent. In Italy, Greece, and Spain cable subscribership is negligible.

Outside Western Europe, the United States, and Canada, significant cable penetration exists only in limited areas. There are over 100 cable systems with more than 500,000 subscribers in Mexico City, for example. In Japan penetration has reached 11 percent, representing roughly 4.5 million subscribers, most of which are served by thousands of small systems outside major urban areas. In 1988 Hong Kong was still in the process of awarding its first cable franchise. There is no cable to speak of in China, India, or Australia.

In summary, including the United States only seven countries have cable penetration exceeding 50 percent. It is probably safe to say that at present nowhere but in the United States do so many viewers enjoy such a large number of programming services via cable. Likewise, U.S. cable operators, with the ability to set their own rates within a liberal regulatory system, are probably the world's least regulated. However, over the next ten years this picture will change considerably as national PTTs loosen their monopoly on TV and major cities worldwide complete the upgrading or construction of cable systems. The greater channel capacity of new systems combined with international program distribution by satellite will help transform cable into a worldwide phenomenon.

Notes

1. "Second Report and Order," 2 FCC2d 725 (1966).
2. United States v. Southwestern Cable, 392 U.S. 157 (1968).
3. "Third Report and Order," 36 FCC2d 143 (1972).
4. HBO v. FCC, 567 F.2d 9 (1977).
5. FCC v. Midwest Video Corp., 440 US 689 (1979) (access channels); Malrite TV v. FCC, 652 F.2d 1140 (2d Cir. 1981) (distant signals).
6. "Must-Carry Rules Fall to Court Edict," *Broadcasting* (22 July 1985): 31. A more liberal version of the rules was later struck down on the same grounds. See "Court Strikes Must-Carry," *Broadcasting* (14 December 1987): 39.
7. *Cable Television Facts* (New York: Cabletelevision Advertising Bureau, 1988): 10.
8. "Cable 88: Fiscally Upbeat, Governmentally Concerned," *Broadcasting* (2 May 1988): 35.

9. Joseph Vitale, "Coveting Thy Neighbor's System," *Channels* (December 1987): 51.
10. Cellia Capuzzi, "Cable Operators: A Rosy Future, But Proceed with Caution," *Channels* (December 1987): 100.
11. Title VI, Communications Act of 1934, as amended by the Cable Communication Policy Act of 1984, Public Law 98-549, 30 October 1984.
12. Delaware, Rhode Island, Vermont, Hawaii, Alaska, and Connecticut preempt local regulation; Massachusetts, New York, and New Jersey supervise it.
13. For a summary of this and other SMATV milestone cases, see W. James MacNaughton, "A Legal Primer for Private Cable," *Private Cable* (February 1985): 26.
14. Michael Burgi, "SMATV Keeps Its Niche," *Channels* (December 1987): 104.
15. Cable data used here come from a variety of sources, including John Tydeman and Ellen Jakes Kelm, *New Media in Europe: Satellites, Cable, VCRs and Videotex* (New York: McGraw-Hill, 1986); Hans J. Kleinsteuber et al., eds., *Electronic Media and Politics in Western Europe* (New York: Campus Verlag, 1986).

12

Cable System Operation and Programming

After obtaining a franchise to operate, all cable systems face two basic challenges: construction and operation of an expensive and complex physical plant, and development of a profitable subscribership through the sale of programming and other services. This chapter first explores the technical side of a typical system and then provides an introduction to options and constraints in system programming.

The Cable System Physical Plant

The signal that reaches a cable subscriber's TV set has completed a remarkable trip through miles of cable, amplifiers, and other black boxes, and often been bounced off a satellite 22,000 miles in space. The equipment that makes this journey possible comprises four major systems: the program supply system, the headend, the signal distribution system, and the subscriber drop.

Program Supply System. Anything seen on a cable system is either originated locally or is relayed in some manner from a source outside the system. Local origination (LO)—the production of local talk shows or news, for example—requires the use of a television studio or mobile equipment. The program is then inserted, either live or on videotape, into the cable system for distribution. All facilities used for LO are part of the cable system's program supply system.

However, most cable programming is imported, not originated by the cable system, and three types of reception equipment are involved. TV receiving antennas, similar to those used on the rooftops of homes, are used to pick up the signals of broadcasting stations in the cable system's vicinity. Familiar dishes, satellite receiving antennas, pick up the program services distributed by satellite to cable system affiliates, including pay movie services, advertiser-supported networks, and superstations. Finally, smaller microwave receiving antennas used for point-to-point terrestrial communications are sometimes used to import the signals of radio and TV stations out of range of conventional antennas and not available by satellite. Both types of program supply equipment (local production and signal importation) feed their signals to the headend.

Headend. Before program signals can be inserted onto the cable, they are brought together and processed electronically at the headend. In pioneer cable systems the headend was simply a shed at the foot of the TV antenna tower where the TV signals were amplified and connected to the cable. Today its function is basically the same, but the headend has become much more sophisticated electronically. Racks of amplifiers and other electronic components—one rack for each cable channel—adjust the power level of each signal, filter it to remove various kinds of noise or interference, and shift it to a frequency corresponding to the cable channel it will be carried on.

Often the headend is located in the same building housing the cable system's TV studios and administrative offices. In some large urban systems, studios and offices may be in separate buildings, even located in different parts of the city from the technical headend, to make them more accessible to subscribers and local advertisers. Regardless of where these other activities are located, the headend is the place where signals are assembled, processed, and inserted at the originating point of the cable distribution system.

Distribution System. The wires that carry the TV signal from the headend to the subscriber are referred to collectively as the distribution system. This is the most expensive part of a cable system, representing perhaps 90 percent of the operator's total investment. After initial construction is completed, the design, construction, and maintenance of the distribution system are ongoing concerns in most cable systems because of system growth and the ever-present problems of equipment damage, deterioration, and obsolescence.

From the headend the TV signals enter the largest-diameter cable in the distribution system, called the trunk (see Figure 12–1). The trunk

Figure 12–1. Basic Cable Television System

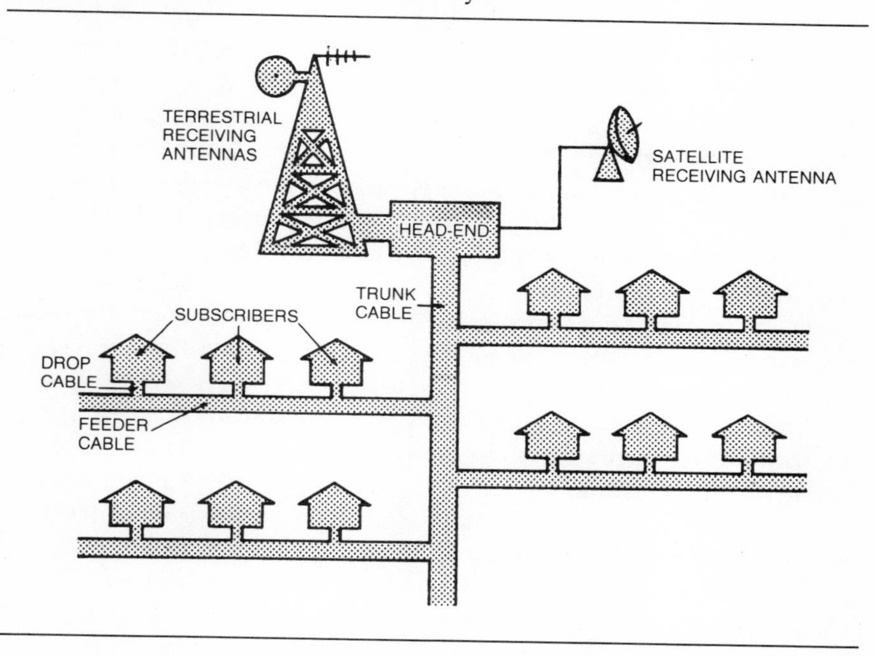

cable, an inch or more in diameter, serves only to transport the signals out into the service areas; no subscriber gets a signal directly from the trunk itself. Trunk lines may extend in more than one direction from the headend. Also, microwave transmission links can be used as a kind of trunk line. Some systems transmit cable signals by microwave from the primary headend to one or more mini-headends, called hubs, each of which has its own service area and distribution system.

Having been transported by trunk line to the service area, the TV signals are split off into smaller cables known as feeder cable, a half inch to one inch in diameter. Most of the cable visible hanging on utility poles is feeder cable, which makes up about 90 percent of the cable in a typical system. The feeder cable transports the signals down each street, through the neighborhoods and residential areas the system serves. At each point where the feeder cable passes a subscriber home, the signal is split off a final time into an even smaller diameter cable, called a drop line, about a quarter to a half inch in diameter. Entering the house or apartment, the drop line is connected directly to the TV set or to a cable converter box.

As the cable signal travels from the headend to the subscriber, it weakens, somewhat like the water in a city water system loses pressure from the treatment plant to the home. Just as a water system must install pumping stations along the way to maintain water pressure, the cable system installs amplifiers on the trunk and feeder lines to boost the signal strength and maintain picture quality. Amplifiers can boost the strength of a TV signal and filter out interference, but they cannot replace video and audio that has been lost. At some point the reamplified signal will have noticeable "snow" in it or will be distorted to an extent that the picture is no longer acceptable to subscribers. Consequently there are limits to how far from the headend the trunk and feeder cable can extend, even though the signal is amplified along the way. This is one reason that hubs are used in cable systems covering large areas.

Amplifiers on the trunk line are referred to as trunk stations. A typical thirty-five-channel cable system usually has about three trunk stations per mile of trunk. Where the signal is split off to a feeder line, a bridger amplifier boosts the signal before it enters the feeder cable. Customer drop lines split off the feeder cable via a device known as a multitap, a sealed metal case with connectors for four or more drop lines. In above-ground cable systems, drop lines can be seen stretching from utility poles into individual homes.

Usually the drop line is attached to a converter, which in turn is connected to the TV set. The converter is used to select the cable channels desired, often using wireless remote control. The need for a converter is a sore point with many owners of "cable ready" TV sets. Although some cable systems use scrambling technology that is compatible with such TVs, most do not. Without the converter, the TV set has no way to decode any scrambled channels the subscriber has paid for. The problem stems from a lack of standardization between the TV set manufacturers and the cable industry. Several types of scrambling systems are in use, and the extent of "cable readiness" also varies with different lines of TVs.

System Construction

Many a would-be cable subscriber has waited impatiently for service to reach his or her home, wondering why it seemed to take so long. However, the design and installation of even a fairly small cable system can be a complex undertaking; where terrain or urban development or system size pose added problems, construction often proceeds only at a crawl. While construction costs in rural areas can be less than $10,000 per mile of cable, they can run to ten times that amount in urban systems where a lot of underground installation is involved.

System construction begins with making maps of the areas to be served by the system, including every utility pole or underground utility duct the system will use. Using these maps the system is designed, usually by engineering consultants using special computer programs. System design requires complex engineering calculations involving a multitude of variables, including placement and length of the trunk lines and feeder system, amplifier power and type, signal strength, headend location, provisions for future expansion, and even weather. Next, before any cable can be hung, every pole to be used must be inspected. In some cases poles need replacement or the utility lines already there need to be moved to make space for the cable lines.

Installation of the cable itself is done by special machines that string a strand of steel support wire between poles and then lash the cable to the strand. Amplifiers are installed on poles, the cable attached to them, and splices made throughout the system to join the sections of hanging cable. In underground installation, cable is pulled through existing utility ducts or buried directly in the ground. Digging involves special problems including disruption of street traffic, locating and avoiding existing underground utility lines, and securing permission for digging on private property. Cable with special sealants and coverings must be used to protect against underground moisture and corrosion. Finally, as portions of the system are checked out and "balanced" electronically, subscriber drops are installed and service begins.[1]

System Programming Constraints

The programming line-up of a cable system is the result of decisions made by the operator in the face of a number of considerations and limitations from strictly technical to legal, economic, and psychological.

Technical Limitations. The most obvious technical constraint is, of course, channel capacity. Channel capacity of a cable system is determined primarily by the capabilities of the amplifiers, rather than the cable itself. Early cable systems were limited to three or four channels, in part because of the size and power consumption of early amplifiers. Solid-state electronics made possible twelve-channel systems in the 1960s. By the early 1970s twenty-channel systems appeared, following the introduction of integrated circuits in cable amplifiers. By the end of the 1970s, thirty-five- and fifty-channel systems were being constructed. Today systems with thirty-five or more channels are common and some have fifty or even 100 channels. However, many twenty-channel systems remain, and twelve-channel systems are still found in less populated areas.

Channel capacity determines not only the number but, to some extent, the kinds of programs available on a cable system. For example, because of public demand for local access channels, local broadcast stations, and advertiser-supported satellite networks, most twelve-channel systems carry only one pay movie service.

Technical considerations also limit the operator's placement of programming on the system. For example, local television stations are not transmitted on the same cable channel as their broadcast channel; that is, a local TV station broadcasting on channel 5 will not appear on channel 5 on the cable converter. Even with the antenna removed, a subscriber's TV set would tend to pick up enough of the broadcast signal to interfere with the cable signal, if they were both on the same channel, causing ghosts.

Psychological Considerations. Subscriber psychology plays a role in channel use also. Industry experience and research suggest that viewers tend to favor the lower end of the converter dial when selecting programming. Programmers prefer the low end and refer to the high end as the "cable ghetto." Television stations claim that if they are moved from a low-numbered spot to a higher one (from channel 4 to channel 50, for instance), their ratings suffer. A cable operator may wish to position a cable network service in a low-numbered channel to enhance viewership, only to face complaints from subscribers and the TV station being forced to vacate. Some consultants suggest that really popular programming should be spread out across the dial, so subscribers catch a glimpse of other services when they search for their favorites.

Federal Laws. Under U.S. copyright law, every cable system pays a percentage of its revenues twice annually into a fund administered by the federal copyright office. This payment entitles cable systems to carry local and distant broadcast signals without having to pay each station individually for that right. This arrangement is called compulsory licensing and is a source of controversy between broadcasters and the cable industry.

The amount of money paid by a cable system is determined in part by the number of distant signals, or superstations, carried. In recent years increases in the rates charged by the copyright office have caused cable operators to reduce the number of superstations carried. Copyright fees have thus become a consideration when deciding how to program a system.

Furthermore, any major change in copyright law could have major repercussions on system programming. For instance, if the present compulsory licensing arrangement were scuttled, cable operators would have to negotiate separately with each television station they carry the amount of copyright royalty to be paid that station. Presumably, popular stations would be in a much stronger position than now to bargain for payment and channel placement. Marginal stations might be dropped.

The Communications Act of 1934 requires that all cable systems with thirty-six or more channels make available commercial leased-access channels.[2] Use of these channels is available to anyone paying a fee established in the cable franchise. In effect, any thirty-six channel system (or larger) does not have full control over programming at least one of its channels. The Act also requires that cable systems electing to carry adult programming make available lockboxes that permit subscribers to control the viewing of that material by minors. The administrative overhead involved with renting or selling lockboxes may affect an operator's decision to carry such programming.

FCC Rules. From 1965 to 1985 the FCC required cable systems to carry "significantly viewed" local TV stations, which in most cases meant all local stations. The effect of the rules was to take a significant number of channels out of the control of the cable operator, depending on the number of TV stations in the market. In 1985 these must-carry rules were held by the U.S. Court of Appeals to violate cable operators' first amendment rights. A more liberal version of the rules was also struck down by the same court in 1987.[3]

Part of the court's reasoning was that the FCC had not produced adequate evidence to justify the rules. At the time of this writing the FCC was collecting data on cable carriage of broadcast stations with the aim of justifying some new version of must-carry. Due to the possible reappearance of some form of must-carry rules, most cable systems have not made major changes in their local station carriage. Even if the threat of new rules diminishes, most systems are likely to continue to carry major local stations because their subscribers demand them. Marginally popular stations, however, may be dropped, giving operators more programming flexibility, particularly in systems with twenty or fewer channels.

Another FCC rule affecting cable programming is the network exclusivity rule. Under it, cable operators are prohibited from bringing in signals of distant TV stations that duplicate the network programming of local stations. For example, if there is an NBC affiliate in the cable system's market, the cable system may not import the signal of another NBC affiliate, unless it blacks out all duplicate programming. In a community with a full complement of network affiliates the network exclusivity rule guarantees that any distant TV signals imported by the cable system will be independent stations. Consequently these rules helped create the initial demand for independent satellite superstations like Atlanta's WTBS.

Lastly, cable programming is affected by the FCC's syndicated exclusivity rules. The "syndex" rules protect local broadcaster's rights to carry syndicated programs in their markets. Usually these are the familiar network reruns like "M*A*S*H" or "The Cosby Show" seen in the late afternoon or early evening. When a television station purchases such programs it pays for exclusivity—that is, the right to be the only station airing a show in a given market. The syndex rules prevent a cable system from transmitting a syndicated program from a distant station if the same program is also broadcast by a local TV station. The cable system must black out the program or drop the distant station. Thus the syndex rules enter into a cable operator's decision about which distant stations to carry, if the syndicated programs on those stations duplicate those of local broadcasters.

Franchise Obligations. Federal law and FCC rules place certain limitations on what sorts of programming requirements are allowable in a cable franchise agreement. For example, franchises cannot require specific programs but can specify categories of programs, such as pay movie channels, or children's programming.[4] Because a franchise is a legal contract to provide services, any programming commitment that does not violate federal rules is binding on the cable operator. For instance, no federal rule requires public, governmental, or educational access channels. But if an operator agrees to provide such channels in the franchise, that commitment is binding. Thus a further constraint on an operator's programming options is any franchise obligation that commits channels to particular uses, such as public access.

Economic Constraints. Because the fundamental objective of running a cable system is to make a profit, available channels must be programmed in a way that enhances their contribution to the bottom line. For example, pay-movie services, like HBO and Showtime, typically split with the cable operator the monthly fee charged the subscriber. Thus, not only does the operator have to consider subscriber desires in choosing a pay service, he or she must also consider which service offers the most profitable split and perhaps the lowest retail price.

Most subscribers are not aware that popular advertiser supported network program services like ESPN or CNN are not supplied free to cable systems. The operator pays a carriage fee to carry the service, around ten to twenty cents per month per subscriber. In return, the programming attracts viewers to the cable system, increasing the operator's revenues. Additionally, satellite networks allow cable affiliates to sell local advertising in designated spots called avails every hour, further increasing potential operator revenue. In deciding whether to carry a particular satellite service, even a "free" one, the operator must balance subscriber desires against carriage fees and potential revenue.

Programming: Satellite Networks

Given the constraints already mentioned, the cable operator selects specific program services for the system's line-up. Satellite-distributed programming falls into three broad categories: basic, pay, and pay-per-view.

Basic Programming. In 1988 about fifty national programming services were available to cable operators via satellite that were advertiser-supported or basic services. The term *basic* means these channels are offered to subscribers by the cable operator for no additional charge beyond the basic cable subscription fee.

Perhaps most familiar of the national basic services are the superstations like WTBS, Atlanta; WOR, New York; and WGN, Chicago. Specialized basic networks (and examples) included sports (ESPN); black-oriented and Spanish-language (BET and Galavision); news, weather, and public affairs (CNN, Headline News, Financial News Network, Weather Channel, C-SPAN); cultural and educational (A&E, Discovery, Learning Channel); music (MTV, Nashville Network, Video Hits-1, Country Music Channel); shopping (Home Shopping Network, Video Shopping Channel); children's entertainment (Nickelodeon); and religion (CBN, Inspirational Network, Eternal Word Network).[5] About a dozen regional basic channels were also offered on satellite, featuring area sports, news, and religious programming.

In addition to these services, the operator could choose from satellite distributed audio-only and text services. Audio-only services included two radio superstations offering jazz or classical music, religious programming, and background music. Among the text services were Associated Press and Reuters news wires, sports news, and electronic program guides.

Pay Programming. The second major category of satellite programming is pay networks, which are offered without commercials to subscribers for an additional charge each month, usually $8 to $10 per channel. In

1988 ten national pay networks offered fare such as general movies and specials (HBO, Cinemax, Showtime/The Movie Channel), adult entertainment (Playboy), family/children's programming (Disney Channel, Festival), and arts/cultural (Bravo). Nine regional pay services were offered, eight of which carried sports programming.

Pay-per-View. A third major type of programming available by satellite to cable operators is pay-per-view (PPV). National PPV services offer movies, concerts, and sports events, charging for them on a per-showing basis. In the late 1970s Warner Amex's Qube system in Columbus, Ohio, the nation's first two-way, fully addressable cable system, was the first to offer local PPV service to its subscribers. By 1985 national PPV services began to appear. Today two are predominant: Viewer's Choice, regularly carried twenty-four hours per day on 120 affiliate systems, and Request Television, carried on 160 cable systems.

To offer PPV, a cable system must be equipped with addressable converters, which can be instructed from the headend to turn on (descramble) a PPV channel for subscribers who order programs and remain off (scrambled) for those who don't order. The operator also needs an efficient system for taking PPV orders from subscribers and knowing which converters to activate.

Addressable technology represents a considerable investment to the average cable system operator, involving the purchase of new converters or upgrading of existing ones and installation of new headend equipment. A further investment must also be made in an order-taking and billing system. Although customer service representatives can take orders by phone and activate converters manually, this approach is much too slow for systems of any size and results in jammed phone lines and frustrated would-be viewers. Several kinds of impulse pay-per-view (IPPV) systems are available that allow orders to be placed, activated, and billed automatically. Each has its merits, and an industry debate continues as to which is best.

Consequently, only about 400 cable systems are addressable, although they do represent some 10 million subscribers. Of these subscribers, only about 5 million are offered PPV programs on a regular basis; many addressable systems carry PPV only sporadically. In 1988 PPV profitability stories in some systems and growing availability of automatic ordering systems made impulse PPV a hot topic in the cable industry, with observers predicting the number of addressable households would surpass 24 million by 1990.[6]

Programming: Local TV Stations

As mentioned earlier in this chapter the federal court's dismissal of the FCC's must-carry rules removed, at least temporarily, the federal requirement that cable operators carry local TV stations. However, operators are well aware that clear reception of local broadcasting stations is still a major reason people subscribe to cable and, furthermore, that cable subscribers spend about two-thirds of their viewing time watching local TV station programming.[7] Consequently, local broadcast stations remain essential in any cable system's programming line-up. Maintaining high-quality transmission of local stations is a critical aspect of cable system operation and programming.

Programming: Local Origination

In addition to satellite programming and local broadcasting stations, the third piece of the cable programming puzzle is local origination (LO), programs whose transmission originates at the cable system itself. LO programming can be produced using cable system studio facilities, or it can be produced elsewhere locally and put on the cable system remotely or on videotape. There are three major categories of LO programming: public, educational, and government access (PEG), leased access, and operator originated.

From 1972 to 1979 the FCC required all cable systems to set aside channels for PEG access. Although the FCC's requirement was struck down by federal courts in 1979,[8] access requirements in local franchises were, and still are, considered valid. A public-access channel provides a free outlet for programs produced by any local citizens or community groups who want to make the effort. The cable operator has no control over the channel's content. Rules for the channel's use are set out in the franchise. The cable system makes available TV equipment, training, and studio time. In larger systems support may include a regular staff for public access production and mobile vans for location shooting, all at cable system expense.

Generally, public-access channels have had a somewhat uneven record of success. While they have become a vital showcase for local programming and opinion in many communities, in others they have fallen into disuse. Public-access channels can also be the center of local controversy because of their open-to-all nature and the lack of editorial control by the system operator. For this reason, federal law protects cable operators from legal action or other responsibility for objectionable programming on public-access channels.[9]

Educational and government access channels, like public-access channels, are no longer an FCC requirement but a matter of negotiation between the cable system and the city. Most franchises have some sort of provision for use of the cable system by local schools and municipal government. Educational-access channels carry everything from school board meetings and night classes to class plays and football games. Government channels provide coverage of town meetings, deliberations of city councils and administrative boards, and forums on public-interest issues.

In some cases local educational and government agencies are provided their own separate institutional cable network by the system operator, referred to as an I-net. Other users, like banks or local businesses, may pay a fee to have access to the I-net for data transmission and other kinds of administrative communications.

The second major type of LO programming involves what are called leased-access channels. As mentioned earlier in this chapter, federal law requires all cable systems with thirty-six or more channels to make available channels that can be rented by users for commercial or other purposes. The goal of this policy is to ensure that, regardless of the nature of PEG access agreements in the local franchise, every cable system with thirty-six or more channels has at least one outlet for commercial use. This is intended to prevent a cable system operator from having a monopoly on commercial programming in the local market. Ground rules for the use of the channel and leasing rates are spelled out in the local franchise agreement. Some users produce local programs and attempt to sell commercial time. Examples include video shopping catalogs and classified advertising services.

The third general type of LO is operator-originated programming. In PEG and leased-access channel programming the cable system operator has no direct editorial control over or responsibility for channel content. However, in some systems, operators find it desirable to set aside a channel for originating programs over which they do have editorial control. In many cases programming consists of local news and sports coverage produced by the operator and supported by local advertising sales, in much the same way that local TV stations function. Cable operators have found that a professional looking local news program provides a cable system credibility and community good will, especially where there is no local TV station to cover local news and events. It can even turn a profit.

Rather than get into local program production an operator may decide to lease such a channel to a local radio, TV station, or newspaper or make it a partner in the channel's operation. Regardless of the use it's put to, however, editorial control and responsibility of an operator-controlled channel ultimately rests with the system operator. Consequently most federal regulations applying to broadcast TV program content also apply to operator-controlled channels, including obscenity and political advertising rules.

Beyond Passive Viewing: Two-Way, Interactive Services

Many cable system operators see themselves as essentially being in the TV program distribution business. Some, though, take a broader view of encompassing the many forms of communication that technically can be transmitted over a cable system. While the industry's current prosperity testifies to the fulfillment of cable's promise as an efficient and profitable method of distributing programs to passive viewers, cable's potential for interactive programmng and services has been barely scratched.

This is largely due to the cost of two-way technology. In a two-way cable system the amplifiers are capable of passing signals not only downstream to the subscribers' homes, but upstream from the subscriber to the headend. Two-way systems require bidirectional amplifiers, converters capable of signal transmission as well as reception, and headend computers and other equipment for receiving and acting on the upstream signals. They are engineered to more stringent technical tolerances, are more prone to problems, and require more maintenance than one-way plants.

Perhaps the best known two-way system in the United States has been Warner-Amex's Qube system in Columbus, Ohio. It is not the first two-way system, but its ambitious applications of two-way technology attracted nationwide attention when it began operation in 1977. The Qube system became more of a showcase and laboratory for Warner-Amex, rather than a profit-making venture, however, and many of its two-way services were abandoned eventually because they simply were not profitable. Qube and other two-way systems around the country have shown the industry how two-way cable works; they just haven't demonstrated how to make it profitable. Consequently, very few two-way systems exist. These systems have tried, with varying degrees of success, three broad categories of two-way services: monitoring services, interactive programming, and information retrieval.

Monitoring Services. Monitoring services use a black box connected to the subscriber's converter that reports to the headend the activation of various types of devices, such as smoke detectors, burglar alarms, or medical panic buttons. For example, when a smoke detector activates, it sends a signal to the monitoring circuitry attached to (or inside) the two-way converter. The converter in turn transmits a code to the cable system headend. Receiving the code, the headend computer alerts an operator with the subscriber's address and phone number. The operator calls to verify the emergency and then notifies the fire department as needed. Burglar alarms and medical alert systems for the elderly or infirm work much the same way.

Monitoring devices can also be used to regulate home utilities and appliances, automatically reducing power for heating and cooling when the house is unoccupied via two-way connections between the converter and thermostats or electrical panels. Technical reliability and support are major considerations in the decision to offer any monitoring services. In entertainment services, breakdown is merely a matter of inconvenience. In monitoring services breakdowns or false alarms can result in property loss or an improper response to a life-threatening situation.

Interactive Programming. The Columbus, Ohio, Qube system demonstrated the technical feasibility of several kinds of interactive programs in which the home audience played an active role by asking questions or making decisions using numbers on their converter keypads. Public affairs programs and talk shows featured instantaneous polling of viewers as to their agreement with the opinions being expressed or their rating of a speaker's effectiveness.

Most popular were interactive entertainment shows like Qube's version of "The Dating Game," which let viewers decide which contestants were paired up. Qube's management found, however, that audience interest in interactive programming was difficult to sustain after the novelty wore off and that innovative and profitable ideas were difficult to come by.[10] Consequently, interactive programming is a rarity in U.S. cable.

Information Retrieval and Transactional Services. In these services the headend computer acts merely as a switch or gateway, connecting the subscriber's converter to computers operated by other businesses or agencies, like stores, banks, or electronic databases. Subscribers can then bank at home, order from an electronic merchandise catalog, purchase concert tickets, make reservations, or catch up on the latest news. Generally, because of the lack of two-way cable systems, these kinds of services use telephone lines, not the local cable system, to link subscribers to the necessary computers. Home shopping, videotext, and similar services are discussed in later chapters.

Notes

1. An excellent discussion of system construction and other subjects touched on only briefly in this chapter can be found in Thomas Baldwin and D. Stevens McVoy, *Cable Communication*, 2d ed. (Englewood Cliffs, N.J.: Prentice-Hall, 1988).
2. Section 612, Communications Act of 1934, as amended by the Cable Communication Policy Act of 1984, Public Law 98-549, October 30, 1984.
3. See "Must-Carry Rules Fall to Court Edict," *Broadcasting* (22 July 1985): 31; and "Court Strikes Must-Carry," *Broadcasting* (14 December 1987): 39.
4. Section 624, Communications Act of 1934, as amended by the Cable Communication Policy Act of 1984, Public Law 98-549, October 30, 1984.
5. See current issues of *Cablevision*, "Cable Stats" section, for current lists of all available program services.
6. See, for example, Janet Stilson, "PPV Growth Accelerating," *Electronic Media* (5 October 1988): 3; Roberta Bernstein, "Pay-per-View: The Players Square Off," *View* (1 January 1988): 60.
7. "How Viewers Divide Their Time among Major Programming Sources" in "Cable TV Facts" (New York: Cablevision Advertising Bureau, Inc., 1988), 10.
8. FCC v. Midwest Video Corp., 440 U.S. 689 (1979).
9. Section 611, Communications Act of 1934, as amended by the Cable Communication Policy Act of 1984, Public Law 98-549, October 30, 1984.
10. Margaret Yao, "Two-Way Cable Disappoints Viewers in Columbus, Ohio as Programming Lags," *Wall Street Journal*, 30 September 1981, p. 31.

13
The Telephone System in Transition

Since the addition of data transmission to traditional voice communications, telephone networks have become the indispensable means of transporting the electronic goods and services of the information age. This chapter offers an overview of the transition of the U.S. telephone system over the last hundred years from a relatively simple audio communication medium to a complex computer-controlled network carrying virtually all forms of information. Two main aspects of that evolution—organizational and technological—are summarized, and the link between them is examined.

Bell and the Origins of the U.S. Telephone System

In recent years the organization of the U.S. telephone system has been altered dramatically. The significance and enormity of this transformation can be appreciated most fully with the benefit of some historical perspective, beginning with the invention of the phone and the first phone company.

Between 1875 and 1877, Alexander Graham Bell, a successful Boston speech teacher for the deaf, received four patents for technical improvements on existing telegraph systems. Among them was the telephonic telegraph receiver, which permitted sound transmission over telegraph wires. Bell entered into an agreement with the fathers of two of his pupils to establish the Bell Telephone Company in 1877 to develop the telephone technically and commercially. In 1878 they opened the first U.S. commercial telephone exchange in New Haven, Connecticut, with twenty-one subscribers. It was soon followed by local exchanges in major cities across the United States and Canada, some two dozen by 1879. These local systems were owned and operated by firms that were controlled by the original Bell Telephone Company through stock ownership and patent licensing arrangements.

During the 1880s the continuously expanding company, renamed American Bell Telephone, successfully fought off legal challenges to Bell's original patents. It also overcame efforts by Western Union Telegraph, the dominant telecommunications company of that era, to establish its own phone system using equipment developed by rival inventors. American Bell also established a subsidiary, American Telephone and Telegraph, to construct a long-distance network serving its local exchanges in cities across North America. The nation's first long-distance line was completed in 1884, linking Boston and New York.

In 1890 a corporate reorganization established AT&T as the parent company. AT&T now controlled not only long-distance service and telephone equipment manufacture (through another subsidiary, Western Electric) but all the dominant regional phone companies across the nation. Thus, little more than a decade after the first local telephone exchange opened, the fundamental nature and structure of the U.S. telephone system was complete. It would remain virtually unchanged for nearly a century.[1]

The Bell System's Fifty-Year Honeymoon

That structure came to be known as the Bell system, a government-regulated monopoly owned by AT&T that served 80 percent of the U.S. population. The remaining 20 percent of the market, most sparsely populated rural areas, was served by 1,400 or so independent phone companies. The Bell system was not only horizontally integrated but vertically as well. All technology was researched, manufactured, installed, and operated by AT&T. Customer-premises equipment (telephones) was built only by Western Electric and could be only leased by users, not purchased. No other company's hardware could be attached to the system. Long-distance service was provided by AT&T Long Lines, which interconnected all the local exchanges across the country. Local exchanges were operated by the twenty-two Bell operating companies (BOCs), with names like Southeastern Bell, Pacific Telephone, Ohio Bell, and so forth.

Controlling interest in these companies was held by AT&T. The Bell system virtually *was* the country's phone system.

AT&T's dominance was not an accident: phone systems in the United States and Europe all evolved in a technological and political climate that favored a regulated monopoly over multiple competitors. Because of the expense involved, competition in local or long-distance service was deemed impractical—a waste of resources—especially in light of a general goal of universal service. Also, the development of telephone service was too important to a nation in terms of commercial growth, public safety, and national security to be hampered by the fragmentation and waste that might accompany competition. In Europe and elsewhere telephone service developed as a monopoly run by a government-owned corporation or government agency. In the United States telephone service evolved in the hands of government-regulated private companies.

In 1934 all U.S. interstate telephone service was placed directly under the regulatory control of the newly established Federal Communications Commission. As the phone system's watchdog, the FCC was tasked with ensuring that AT&T and the independents provided universal service at fair and reasonable rates. It was also empowered to counter any abuses the public or the independent phone companies might suffer from the dominant AT&T. Local and intrastate service regulation remained in the hands of state utility commissions.

AT&T became the world's largest business enterprise, with annual revenues exceeding the gross national product of many lesser developed nations. AT&T's Bell Laboratories became the world's largest nongovernmental research organization, responsible for many of the key inventions (including the transistor and the laser) that made possible the computer and modern telecommunications. This arrangement worked to most people's satisfaction for several decades. Subscribers considered the U.S. telephone system the most advanced and reliable in the world. British researcher Jeremy Tunstall has pointed out that in the years after World War II much American business and pleasure travel was in European countries where the expansion and modernization of telephone systems had been set back by the war. Their chaotic condition further increased Americans' conviction that the Bell system was inherently superior.[2]

The Honeymoon Comes to an End

Despite the general satisfaction of the average subscriber, the postwar years brought the first signs that AT&T's long honeymoon with regulators and corporate customers was over. In 1949 the Justice Department filed suit against AT&T claiming, among other things, that the exclusive use of Western Electric as an equipment supplier artificially increased expenses and raised telephone rates. In the 1956 settlement of that suit AT&T was allowed to continue the Western Electric arrangement but had to agree to stay out of manufacturing and sale of new telecommunications products and services, such as computers.

At the same time the emergence of practical business computers and other advances in microelectronic technology began to change the needs of commercial telephone customers. The decreasing cost of some electronic components and the invention of new technologies gave rise to new telecommunication products from upstart competitors. Customers wanted to use the phone system to link computers in separate locations; some wanted to attach new kinds of non-AT&T equipment to their phone lines. Others wanted to build their own microwave communication systems, circumventing the Bell system altogether.

Understandably, AT&T resisted attempts to attach any non-Bell equipment to "the system." It questioned the quality of foreign equipment and refused to permit the attachment of even the simplest devices on the grounds that unforeseen damage might result. Generally, when it came to the demand for new transmission services, such as computer lines or other innovations, AT&T took the position that it knew what was best for the U.S. telephone system (it had invented it, after all) and it would make necessary changes and improvements as they were deemed appropriate, using its own equipment and research facilities—the world's best. The U.S. government, however, disagreed and beginning in 1956 started to undermine AT&T's monopoly position.

Three Decades of Deregulation

In its 1956 "Hush-a-Phone" decision the FCC ordered AT&T to permit customers to attach nonelectronic devices to their phone sets as long as no damage was done to the telephone system. The Hush-a-Phone was a simple foam collar attached to the mouthpiece to muffle conversation. This decision signaled the end of an era: never before had customers been allowed to attach anything to telephone company equipment. It also set a precedent for the dramatic deregulation of the telecommunications industry over the next three decades. The deregulatory dam broke with the FCC's 1968 Carterfone decision, which extended the Hush-a-Phone ruling by allowing the attachment of non-AT&T electronic devices (answering machines, switchboards, fax machines, and so forth) to the system by customers.

Today customers may attach any equipment they desire to their phone lines, as long as that equipment has been approved by the FCC. Competition among numerous suppliers has led to declining prices in customer-premises equipment of all kinds and a wide variety of speciality phones and other intelligent terminal equipment with ever-increasing features.

During the last twenty years, deregulation has opened other aspects of the telecommunications industry to competition as well. FCC and federal court actions have allowed new long-distance carriers like MCI and Sprint to compete with AT&T, leading to a greater variety of long-distance options. Furthermore, subscribers have the option of establishing their own private-line long-distance systems. The FCC has also opened up the area of data transmission so many companies can use the telephone system to provide networking services linking customer's computers, access to computer databases and electronic messaging, and other services.

Restructuring AT&T

While the FCC was opening up the telephone system to competition in long distance, customer equipment, and data transmission, a separate action was taken by another government agency that would have a profound impact on today's telecommunication industry. In 1974 the Justice Department filed an antitrust suit against AT&T, with more far-reaching allegations than the 1949 action mentioned earlier. Justice's central complaint was that the interlocking organization of the Bell system gave AT&T an unfair advantage over any outsiders who ventured to compete with it. That is, one part of the company could divert profits to another, allowing it to artificially reduce prices when faced with competition in any one segment of its operations.

For example, AT&T was accused of attempting to eliminate MCI competition on some long-distance routes by reducing AT&T rates on those routes below the level that operational costs would normally permit. It could then make up the difference by raising rates on other routes where there was no competitor or by borrowing money from other areas, such as equipment leasing or charges for local service. Generally, the Justice Department argued that telephone customers would be better served and true competition could take place only if AT&T's long-distance, local service, and equipment manufacturing activities had to stand on their own. AT&T argued that in many instances this sort of cross-subsidization of services worked to the subscriber's benefit and produced data to show, among other things, that long-distance profits helped keep down the cost of local service.

Table 13–1. Regional Bell Holding Companies

Holding Company	*Major Bell Operating Companies It Controls*
Pacific Telesis	Pacific Telephone, Nevada Bell
U.S. West	Pacific Northwest Bell, Mountain Bell, northwestern parts of old Southwestern Bell
Southwestern Bell	remaining, original Southwestern Bell
Ameritech	Wisconsin Telephone, Illinois Bell, Indiana Bell, Ohio Bell, Michigan Bell
Bell South	South Central Bell, Southern Bell
Bell Atlantic	C&P Telephone Company, Bell of Pennsylvania, Diamond State Telephone, New Jersey Bell
Nynex	New York Telephone, New England Telephone

As was the case with the first suit in 1956, the second AT&T antitrust suit never went to trial; it was settled out of court. In 1982, after seven years of legal maneuvering, the Department of Justice and AT&T agreed to a legally binding settlement that essentially did away with the system begun by Alexander Graham Bell a hundred years earlier. The 471-page agreement committed AT&T to the largest corporate restructuring in history.

All twenty-two BOCs, the corporate descendants of the local companies set up in the 1880s, were split off from AT&T and reorganized into seven independent regional Bell holding companies (RBHCs) (see Table 13–1). Though AT&T retained its long-distance, manufacturing, and research subsidiaries, it had given up the largest part of its organization, approximately $100 billion in assets, and shrunk to only one-fifth its former size.

In return, the new "lean and mean" AT&T was permitted to do things it could not do under the first antitrust suit settlement: enter the fields of information processing, computer manufacturing, and new transmission services in addition to its traditional long-distance service and telephone equipment manufacturing. The only restriction on AT&T's new competitive freedom was a seven-year ban on electronic publishing. There are also regulatory limits on the extent AT&T can subsidize its new unregulated undertakings, like computer manufacturing, with profits from its regulated activities, such as long-distance service, in which it still controls about 80 percent of the market.

The Phone System Today

The result of thirty years of liberalization and divestiture in the telephone industry has been a shift from a generally predictable, easy-to-comprehend, monopoly structure to a multilayered, complex, competitive structure, replete with the consumer opportunities and pitfalls of any open marketplace. Your telephone equipment, once built only by Western Electric, may now be manufactured by any of dozens of companies in and outside the United States and can vary dramatically in price and quality. Where once it only could be leased from an AT&T-owned local phone company, it now can be purchased at retail phone outlets and discount electronics stores. Although your local phone company still is the only one in town providing local service, it now can connect you not only to AT&T for long distance but to a number of national and regional carriers who offer varying rate plans and qualities of service. Companies offering new kinds of telecommunication services and equipment now find it easy to connect them to existing local and long-distance systems. Organizations desiring to bypass existing local or long-distance companies may do so by leasing or building their own private communication systems using microwave and satellite transmission systems.

The BOCs, once limited to providing only transmission service, now are diversifying into equipment sales (not manufacture) and a variety of new telecommunication services (such as voice messaging), limited only to the extent that they cannot originate or control the content of information they transmit. The largest RBHCs are now as big as AT&T itself, in terms of revenues, and may become significant competitors in some of the new markets into which AT&T is venturing.

The Telephone System's Technological Transition

The sweeping changes made by regulators in the organization of the U.S. telephone system have been spurred to a large extent by dramatic technological advances. The rapid development of new technologies, beginning in the late 1940s, soon made the old organizational scheme for the industry obsolete.

Technological change had always been a part of the telecommunications business, but it was change that was instigated and controlled mostly by AT&T. It was incremental and could be accommodated by the existing system. The new changes were different. The development of long-distance service provides a key example.

In 1893, with the completion of the nation's longest long-distance circuit, 1,200 miles from Boston to Chicago, it appeared that the limits of existing technology had been reached. There was no effective way to amplify a signal over a longer distance, due to the lack of electronic amplifiers to boost the signal at points along the line. But the development of the vacuum tube some years later made electronic amplification possible, and in 1915 the first coast-to-coast phone calls were placed.

For the next thirty years, national long-distance technology consisted of refinements of this system of amplified wire lines. In the late 1940s, based on wartime research advances, AT&T began construction of microwave transmission systems, greatly improving the quality and capacity of the long-distance network. The advent of microwave transmission systems heralded a new era in telecommunications technology, a significant break with incremental improvements in the past. Microwave system technology, unlike wire transmission systems, permitted some companies to build affordable private long-distance networks, sidestepping the Bell system. For the first time, new technology had provided an economical alternative to using AT&T's facilities. Correspondingly, the FCC set aside microwave frequencies for private transmission systems in 1959, then ten years later opened the regulatory door for rival companies such as MCI and GTE to provide public long-distance service in direct competition with AT&T.

Compounding the destabilizing effect of terrestrial microwave technology on the long-distance industry was the development of communication satellites. The geosynchronous communication satellites launched in the early 1960s were nothing less than the ultimate microwave repeater stations, even using the same frequency bands already assigned to terrestrial systems. They made it possible for anyone with access to sending and receiving dishes to construct their own transcontinental, even international, long-distance system. This effectively eliminated AT&T's technological monopoly on long-distance service.

The link between technological change and industry change is also exemplified in the development of the transistor and subsequent advances in solid-state and microelectronics. In broad terms the invention of the transistor gave impetus to the modern computer industry by making possible a more affordable, powerful, and reliable computer than had been the limit of vacuum tube technology. The spread of computers into private industry gave rise to demands for economical data transmission service. At the same time the development of microelectronics in the 1960s caused a dramatic increase in the capability and a simultaneous reduction in the cost of virtually all kinds of electronic equipment. This led to the production of new, "smarter" kinds of customer telephone equipment, in many cases by companies involved in computer manufacturing. Consequently, beginning with the Carterfone decision mentioned earlier in this chapter, the FCC effectively ended AT&T's monopoly on phone equipment manufacture by allowing the attachment of rival products to Bell system lines.

It is clear, then, that the telephone system transition in the United States has been accelerated by dramatic technological advances that made it possible for regulators to shift from a promonopoly position supportive of AT&T and the Bell system to a philosophy favoring competition among multiple manufacturers and service providers. As AT&T officials have pointed out, these changes in the telephone system's structure hold a powerful irony for AT&T.[3] It was AT&T's Bell Laboratories that invented and developed microwave transmission technology, the transistor, and many aspects of microelectronics. The first active communications satellite to be orbited, Telstar in 1962, was designed and built by Bell Labs. Thus, through Bell Labs, AT&T gave birth to much of the technological change inspiring the regulatory shifts that ultimately led to the dismantling of AT&T itself with the 1982 divestiture agreement.

The System of the Future: Fiber Optics, ISDN

In the United States and most developed countries the traditional voice communications system has spawned a data transmission system that has evolved to meet the mushrooming demand for high-speed business computer communications. At the same time many new types of customer terminals, such as fax machines or personal computers, are using the voice network for data communications. In the near future we can expect dramatic increases in the capacity of both types of networks as fiber optic technology joins and even replaces today's conventional wire cables. Further down the road it is possible that both systems, voice and data, will merge into a single unified telecommunications network that will allow subscribers to transmit and receive voice, data, and video through the same plug in the wall.

Fiber Optic Transmission. Fiber optic cable and associated devices allow voice, video, and data to be transmitted by a beam of laser light rather than on a continuous wave of electrical energy. Optical transmission has been made possible by the development of the laser and processes for manufacturing optical fiber—flexible strands of glass. There are two main types of optical fiber: a low-grade fiber, used to carry conventional light in applications such as automobile dashboards, novelty lamps, and signs, and a high-grade fiber, which can transmit laser light without distortion. This communications-grade fiber was first commercially produced in the United States in the early 1970s by Corning Glass Company.

In a fiber optic transmission system, the voice or data signal is converted from electrical form into a series of pulses of laser light. These pulses are an optical version of the on/off electrical impulses of digital code. The unique intensity and purity of laser light make it the ideal light source for carrying digitally encoded communications over fiber optic lines. At the receiving end the pulsating laser light beam is captured by a photosensitive cell and converted again into its original electrical form.

Fiber optic transmission has several major advantages over other technologies. Because it is not electrical—that is, there is only light and not electricity flowing through the cable, and the cable itself will not conduct electricity—optical transmission is not affected by electrical or radio frequency interference or even lightning. This immunity from most kinds of interference allows optical transmission systems to operate in technically hostile environments with greater reliability than other media. Fiber optic systems are also much more secure because they do not "leak" electrical energy like conventional transmission systems and cannot be physically tapped without detection.

Most significantly, fiber optic systems have a much greater information capacity than conventional transmission systems. A fiber optic cable can carry many times the voice, data, and video that can be squeezed into a similar diameter conventional cable. Thus, converting to optical fiber allows telephone systems to dramatically increase their capacity without expanding underground ducts or adding more cables to existing telephone poles.

Fiber optic transmission systems can even be more advantageous than terrestrial microwave or satellite communication links, depending on the specific application. While they are generally cheaper, both terrestrial and satellite microwave systems are subject to interference from precipitation or frequency crowding and are limited to line-of-sight transmission paths. In the late 1980s satellite transmission was cheaper than fiber optic, but that economy depended on the supply of communication satellites, which can fluctuate. Setbacks in satellite launches and a general trend of increasing demand for transponders can reduce that edge.

For all these reasons, telephone systems around the world have been upgrading portions of their existing networks with fiber optic links since the late 1970s; about 7 million miles of cable had been put into service worldwide by 1987.[4] Dozens of countries have undertaken ambitious projects to link major urban centers with fiber optic transmission systems. In the United States high-traffic telecommunication corridors such as Boston–Washington and San Francisco–Los Angeles were among the first to receive fiber optic service. In France, where the Minitel videotext service has experienced dramatic growth (see Chapters 14 and 15), fiber optic systems are being installed not only on long-distance lines but directly into the home to bring Minitel and future information services to subscribers in some prototype projects. With the exception of France, fiber optic technology was being applied primarily to long-distance service routes in the late 1980s. Standardization and technical problems have slowed the conversion to fiber optics of local phone systems ("local loops") and private in-house networks, which remain wedded to conventional transmission methods.

The ISDN. The international installation of fiber optic long-distance circuits represents the first step in the long-term development of a new-generation global telephone system, a system based on digital technology rather than conventional transmission techniques. The current upgrading of long-distance lines, which has been underway for over a decade, actually is the simplest part of this transformation. The conversion of local telephone systems and, eventually, individual subscribers' equipment is a much more complex problem.

The Integrated Services Digital Network (ISDN) is an international master plan for replacing present transmission services with a fully integrated, streamlined digital system. When fully implemented, the ISDN will allow audio, video, and data to move from one place to another—even one country to another—over the same lines. There will no longer be a need for a separate system for data transmission. A typical phone line on the ISDN will have more than fifteen times today's capacity, allowing it to carry audio, video, and high-speed data simultaneously. Furthermore, all customer transmission equipment, from telephones and fax machines to computers and video equipment, will be able to plug into the ISDN using the same standardized wall jack.

Although AT&T, the BOCs, and representatives of many of the world's other telephone systems have already been at work on developing a compatible set of international standards for the ISDN, the expense and complexity of such an undertaking ensure that it will take years to become a reality. Eventually, though, ISDN will make the term *telephone system* obsolete because the phone–voice communication–will be only one of many transmission services the integrated international network will provide. Over the next twenty years, what is today mainly a voice transmission system will increasingly become a true global telecommunication network capable of sending and receiving almost any kind of information from your home or office to any other subscriber, all through one plug.

Notes

1. A concise summary of the invention of the telephone and the early expansion of the phone company can be found in John R. Bittner, *Broadcasting and Telecommunication,* 2d ed. (Englewood Cliffs, N.J.: Prentice-Hall, 1985), 40–50.
2. Jeremy Tunstall, *Communications Deregulation: The Unleashing of America's Communication Industry* (London: Basil Blackwell, 1986), 91.
3. For an account of divestiture as seen by AT&T management, see W. Brooke Tunstall, *Disconnecting Parties* (New York: McGraw-Hill, 1985).
4. "Fiber Optics," Report #IT35-100 (Delran, N.J.: Datapro Research, 1987): 101.

PART V
Electronic Publishing and Retailing

14
Instant Information: Videotext and Teletext

Over twenty years ago, when specialized computer data bases were being developed for use by scientific research organizations, some futurists and businesses looked to a day when computerized electronic information services would be available on a mass scale, reaching millions of average consumers. This chapter surveys efforts to develop such systems in North America and elsewhere and outlines the main transmission modes available.

Kinds of Electronic Publication

The term *electronic publishing* refers to the great variety of ways that information, usually text and graphics, can be distributed in electronic form on a mass scale using new technologies. Examples of electronic publishing include computer software, encyclopedias on compact disc, news wire services and financial market data on cable, and so forth.

These and other examples fall into two major categories according to how the publisher disseminates the information: offline and online electronic publication.[1] In offline publishing the information in its entirety passes into the possession of the purchaser or user, who then may access it an unlimited number of times for no further charge. For example, computer software or reference materials on compact disc reach the user in a physically tangible and transportable form, an optical disc or a floppy magnetic disk. They are then accessed repeatedly using an electronic reader like a personal computer or compact disc player. In contrast, with online electronic publication, the information remains in a publisher-controlled computer, usually accessible only on payment of a fee for each use.

Both online and offline information can be searched or browsed by the user, with telephone line charges usually incurred in online searching. A major difference between online and offline publication is that online information can be continuously updated, refined, and otherwise managed. In offline electronic publishing the publisher has no way to immediately expand or update the database once it has been sold. New editions or supplements must be distributed to each user.

A second major characteristic of different forms of electronic publication is the level of accessibility or interactivity. All offline publications can be searched or browsed by the user, limited only by the speed of the storage medium and the friendliness of the system. Online information can be equally interactive, but this may be limited by the cost of phone lines needed to connect users to the database. Online publication may also take a passive form that is not interactive. For example, the AP newswire telecast on cable is online in the sense that it is continuously updated and remains under the publisher's control, but it is passive because it cannot be searched actively by users; they must wait for the news category of interest, like sports, to roll by.

Thus, only interactive online electronic publication has an immediacy or live quality combined with the searchability found in other forms of electronic publication. These two advantages lend themselves especially well to certain kinds of information content, information that has both a short shelf life and features a great deal of detail in many categories, such as stock quotes and airline schedules.

Such services can be transmitted to users' homes or offices by either wire or over the air—systems referred to as videotext and teletext, respectively. As the following sections indicate, they differ in capacity and interactivity, and each has its advantages.

Videotext Transmission Systems

The term *videotext* generally refers to any system for interconnecting a user with a remote computer database by wire in a way that allows the user to search the database actively for information. *Videotext* usually implies complete interactivity, a two-way link that allows the user to query the database for any part of the information it contains.

The most common way to connect the user to the database is by phone lines. The phone line is connected to the user's personal computer or an ordinary TV set equipped with a special videotext terminal and keyboard. Using either the personal computer keyboard or the keyboard attached to the TV set, the subscriber selects desired information by typing appropriate commands according to a series of menus and instructions provided by the database computer. One screenful of information (text and/or graphics) is known as a frame or page.

A less common way to connect a user to a videotext database is via a two-way cable system. In this arrangement the database is connected by phone lines to the cable system headend computer, which passes on the information to the subscriber. The subscriber uses the keypad that comes with the two-way cable terminal to type instructions and select information. From the subscriber's standpoint, both the phone line and cable methods of providing videotext work essentially the same; the same information is transmitted using different systems.

Because only a small percentage of U.S. cable systems are two-way, videotext via cable is rare; most videotext subscribers are connected to their services by phone lines. Furthermore, most of these use personal computers as videotext terminals. Specialized (known as dedicated) videotext terminals for use with home TV sets can produce higher-quality color graphics but cost as much or more than basic home computers and are not as versatile. Consequently, they have not generated consumer interest in U.S. pilot projects, despite major efforts to persuade people to buy or rent them, as described below. Thus, although alternative transmission and reception systems exist, nearly all U.S. videotext subscribers are personal computer owners who pay long-distance charges to connect to databases by phone lines. With personal computers in most workplaces and about 20 percent of U.S. households, the potential market for videotext is limited but still quite large. Development of inexpensive dedicated videotext terminals would expand that potential market technically to virtually every home in the country.

Finally, two different approaches can be used in designing a videotext system: centralized or decentralized. In a centralized videotext system all information services must be either stored in the system's computer or routed through that computer, as if it were a gateway. In a decentralized system information providers maintain their own database computers, which users access separately. In a decentralized system the only commonality is the phone system linking all the users and information providers.

World Videotext Systems

U.K. Prestel. In the early 1970s British Post Office researchers experimenting with video telephone technology developed a method for connecting ordinary home television sets to computers over residential phone lines. Using a keyboard and some decoder circuitry, home receivers could act as computer terminals, allowing a viewer to call up thousands of pages of information on the home screen. The system, called Prestel, was unveiled in 1974. It was the world's first commercial videotext system. Regular public operations began in 1979 with the objective of increasing phone company revenues by providing a new consumer service using mostly existing equipment. By 1988 virtually every telephone in the United Kingdom could access Prestel with a local call, and 70,000 terminals were in use—half commercial, half residential.[2] Prestel originates its own content but also acts as a gateway for other information providers, like newspapers and business data services. Transactional services like home shopping and banking are also available.

France's Teletel. In France a different approach was taken, with considerably different results. The French PTT developed its own version of videotext technology, called Antiope (a Greek goddess), with higher-resolution graphics and more colors than Prestel. Teletel, a videotext service based on the Antiope standard, began in 1981 using a simple terminal called a Minitel. A basic Minitel terminal consists of a small TV screen with a compact keyboard/cover that folds out in front. The entire unit takes up no more space than the average coffee-maker and connects directly to phone lines.

Minitel service started out as an electronic telephone directory, allowing users to locate virtually any telephone subscriber in France by name, street, town, profession, and so forth. Terminals were provided free of charge to encourage use. As other information providers added their databases and services to the system, the free terminals became very much in demand. Users pay the phone company about $10 an hour for system access. Minitel is unique in that it is a decentralized videotext system; information providers maintain their own databases, separately accessible. The phone company makes its money by supplying the common link (the terminals and the data transmission lines), monitoring which services are used, and charging information providers according to the number of times their databases are accessed. All services are billed through the phone company and appear on the subscriber's and information provider's phone bill.

Popular services include message centers similar to computer bulletin boards in the United States and dating services, which permit anonymous conversations between prospective couples. As the number of users continues to increase, transactional services like shopping and banking are becoming popular on Minitel, and print media like newspapers are playing a larger role as information providers.[3] The popularity of Minitel has made France's videotext system the world's largest and most successful. It has demonstrated that free dedicated terminals attract large numbers of consumers who otherwise would not purchase or use them. Minitel terminals are expected to find their way into nearly 3 million French homes and businesses by 1990.

Other European Systems. Shortly after the British system got under way, PTTs (postal and telecommunication ministries) in other European countries launched commercial videotext services based on Prestel technology, beginning with Finland in 1980 and followed by the Netherlands, Sweden, and West Germany. Today most remaining Western European countries have PTT-run videotext systems in various stages of development. The majority serve primarily business users, acting as gateways for commercial information providers. However, consumer use continues to grow. Although most are based on Prestel technolgy, some of the newest systems are based on a new common European videotext standard, CEPT, which offers more sophisticated graphics.

In summary, videotext in Western Europe consists of national variations on the Prestel and newer CEPT systems, plus France's Antiope. Europe—and the world's—largest contingent of videotext users is France's more than 2 million subscribers, with less than a quarter of a million total subscribers in other European countries.[4]

Canada and Japan. In Canada, videotext is administered through the Federal Department of Communications using a system called Telidon that is compatible with France's Antiope standard. When Canada began videotext development in the late 1970s, its aim was to improve the graphics capability of the Prestel system already in place in Britain. The sophisticated Telidon system developed by Canada through government-funded research has been a technical success. Major videotext pilot projects have been cosponsored by the Canadian government and private information providers, including a public-access videotext system, Teleguide, featuring free terminals in shopping malls and hotels in several Canadian and U.S. cities. In 1988 Bell Canada announced a new trial modeled after France's successful Minitel service.[5] But widespread residential consumer videotext has not developed, in part due to changing government funding priorities.

In Japan the videotext system is called CAPTAIN and, as is the case elsewhere, is still in a relatively early stage of commercial development. Japanese videotext had to be designed to accommodate the complex graphics required by an alphabet of several thousand characters. Beginning with trials in 1979, videotext began commercial service four years later and can be found in various major cities, primarily servicing business rather than home subscribers at this stage. Should videotext eventually become a mass medium in China, the CAPTAIN system would be in an excellent position to become the Chinese standard because of its ability to display thousands of complex characters.

U.S. Videotext Trials

The Standards Problem. In the early 1980s, with videotext under government development in Europe, Canada, and Japan, major U.S. media companies began to investigate the commercial potential of videotext. European videotext was, and still is, the child of the PTTs, funded by government. In the United States, where telecommunication service is provided by the private sector, the government has not taken an active role in developing videotext. The primary agency involved in such matters, the Federal Communications Commission, declined to designate a national videotext standard—that is, to select a particular videotext system as the official U.S. technology. Consequently, firms attempting to develop videotext systems in the early 1980s were left on their own in deciding which system to use. Furthermore, because there was, and still is, no government funding for such projects, prospective videotext systems would have to pay their own way eventually through profits.

The standards problem was partially solved by AT&T in 1982. Because it would be the carrier of any videotext system developed, AT&T developed a set of transmission standards for systems planning to use its lines for connections. Called NAPLPS (North American Presentation Level Protocol Syntax), it accommodates the sophisticated graphics and color of a Telidon or Antiope-type system but not the lower-resolution Prestel system. In this way Telidon/Antiope videotext technology became the de facto standard for high-resolution, color videotext in the United States.

Viewtron, Gateway, and Keyfax. Initially, the key to videotext profitability appeared to lie in a mass-market approach—that is, devising systems that would attract millions of subscribers and major national advertisers. Three major mass-market videotext trials were undertaken in the United States, all essentially similar efforts to develop interactive systems with high-level color and graphics that would appeal to advertisers and a mass audience. The first, Viewtron, was launched in the Miami, Florida, area, jointly operated by AT&T and Knight-Ridder, a major newspaper chain. Viewtron trials began in 1983 and regular commercial operation commenced in 1985. Combining AT&T-supplied terminals and computers and Knight-Ridder's information resources, Viewtron offered several thousand pages of news, sports, weather, stock market quotes, restaurant menus, travel guides, movie reviews, games, and other information. Viewtron also acted as a gateway for transactional services provided by other firms, such as home shopping, banking, and airline reservations.

In 1984 the Times-Mirror company, a leading publishing and broadcasting chain, launched a similar service in Los Angeles, also using AT&T-built equipment. Called Gateway, it offered the same kinds of information and transactional features as Viewtron. Shortly afterwards a third major trial, known as Keyfax Interactive Information Service, got underway in Chicago. Funded by the Chicago Sun-Times, a computer manufacturer (Honeywell), and the area telephone company (Centel), Keyfax offered essentially the same kinds of services as Gateway and Viewtron.

By early 1986 all three videotext projects were abandoned, unable to attract enough subscribers to ensure future profitability. Together Gateway and Viewtron had cost their developers about $50 million each. All three experiments appeared to suffer because initially they could only be received by people willing to pay $600 to $700 for a special terminal (a decoder box similar to a cable converter and a small keyboard). Even after prices were slashed, the terminals would not sell; only 3,000 were purchased in Miami, for example. Subsequent efforts to rent terminals were fruitless. To most people they simply seemed to be more trouble to set up and operate than they were worth.

Second, the videotext companies became convinced through their own research that most of the services they were offering simply were of no great interest to the average family, especially after the novelty wore off. The $20 to $30 per month subscription fee compounded consumer resistance.

However, in all three trials, a small group of consumers did respond positively to videotext: computer owners. Viewtron, Gateway, and Keyfax offered software to computer owners that allowed them to receive a "plain vanilla" monochrome version of the services. About 17,000 home computer owners signed up for Viewtron nationally when software became available. But with computers in less than 15 percent of U.S. households at the time, a national mass market for videotext seemed remote. Furthermore, the newspaper chains and other media companies involved had satisfied themselves that videotext was not a threat to their traditional businesses. The projects were shut down.

PC Videotext. One of the handy things a personal or business computer can do is "talk" to another computer over phone lines. Text sent from one computer to another is usually transmitted in a standardized code called ASCII, which is designed to accommodate only text and simple graphics, not high-resolution color like NAPLPS. During the early 1980s while mass-market videotext experiments using dedicated NAPLPS terminals were getting underway, online interactive information services were steadily attracting thousands of computer users across the nation.

Unlike the specialized educational and scientific databases developed in the 1970s, which required trained operators to extract information, these new services were aimed at the average home computer owner and could be accessed through easy-to-use menus. Unlike the failed videotext trials that used high-tech dedicated color terminals, this form of videotext transmitted simple monochrome text and graphics to a variety of commonly owned home computers.

Compuserve, the Source, and the Dow-Jones News Retrieval Service were among the first to become available to PC owners and offered thousands of pages of news and information, electronic messaging (computer bulletin boards), and transactional services like airline reservations. By 1988 these services and more than three dozen others had attracted around a million subscribers.[6] Although high-tech videotext had floundered as a national medium, less glamorous PC-videotext was thriving in its niche market of computer owners.[7]

Next-Generation Videotext. The steady growth of PC videotext and the popularity of France's Minitel system have encouraged companies still interested in mass-scale electronic information services. Taking their cue from the Viewtron, Gateway, and Keyfax failures, however, they have abandoned high-tech, NAPLPS-type videotext with expensive, specialized terminals and are pursuing something more like Minitel or CompuServe.

The most prominent of these new-generation trials is a $250 million joint venture by IBM and Sears, the country's largest retailer, which formed a videotext company called Trintex. Being tested in selected cities in 1988, Trintex's Prodigy service consists of a package of information and interactive services that can be accessed only by using a personal computer. Unlike earlier systems, Prodigy users will pay only a $10 to $15 monthly flat fee for access; there are no additional fees for time on-line and no long-distance charges. Features are reported to include electronic columnists like Jane Fonda and Howard Cosell, who will respond to user's electronic questions or comments, stock quotes, travel reservations, games, and shopping from Sears and other retailers, including grocery stores.

Prodigy is said to be able to target users individually with advertising using demographic information supplied when they subscribe. Advertisers then pay only for specific types of users they know have been exposed to their messages and product information.[8]

At the same time five U.S. regional telephone companies have announced plans to pursue small-scale videotext pilot projects, in which they would presumably play a role somewhat akin to that played by the French PTT in the Minitel system, even using U.S. versions of Minitel terminals.[9] One major difference, however, would be that the phone companies would not offer electronic directories like those whose popularity helped get the Minitel service started. At present U.S. phone companies legally may transmit only videotext services owned by other parties, they cannot offer their own information services. These constraints could, at some future date, be removed.[10]

A third new approach to videotext is public-access terminals or kiosks, which provide specific information or transactional services in certain settings. Free walk-up terminals, like the Canadian Teleguide system, are appearing in shopping malls, stores, airports, and hotels, providing visitors electronic shopping tips, catalog shopping, reservations, information, and even hotel check-out services. Such applications are a far cry from the sort of high-tech residential information utility attempted in trials like Viewtron and Gateway. However, at least one industry analyst has predicted that in all its forms, including PC videotext and public-access terminals, videotext will be a billion-dollar industry by 1990.[11] These kinds of videotext may well lay the groundwork for broader public acceptance and, ultimately, reliance on interactive electronic information systems in the future.

Teletext Transmission Systems

Teletext is a technology that allows conventional TV broadcast stations to transmit screens full of text and graphic information. Using a decoder and keypad, subscribers select teletext pages for viewing on their home TV screens. While videotext is always transmitted by wire, teletext is always transmitted as part of a television signal, either over the air or by cable-TV transmission.

Use of the VBI. The picture on any TV set is composed of horizontal lines—525 lines in the United States and other countries that share the U.S. transmission standard. Not all of these scan lines contain picture elements. Some blank lines must be transmitted to separate the end of one image from the beginning of another. Twenty-one of these blanking lines can be seen as the black band that appears on the screen when the picture is rolled up or down slightly.

Called the vertical blanking interval (VBI), these lines also can be used to transmit information in addition to the picture itself because they are empty and normally out of sight. Teletext transmits text and graphics by inserting them into the TV signal VBI at the television station and then removing them with a home decoder box for display on an unused channel or at the bottom of the regular picture. During this process the home TV receiver continues to function normally. Without a decoder to capture and display the teletext information, a viewer would not even be aware it is being transmitted.

Decoders and Interactivity. A teletext decoder can be built into a TV set at the factory or can be attached like a cable converter and placed on the top of the receiver. A remote-control keypad allows selection and display of text. As with videotext, the first teletext screen provides a table of contents or menu, with instructions for calling up the various categories of information that follow.

In a teletext system all transmission flows in one direction: from the station to the home receiver. Pages are transmitted in a continuous cycle from the station, one after another. The user cannot send any information back to the teletext computer at the station but can only select and view desired pages. However, the one-way nature of the transmission system does not prevent the user from interacting with the information once it has reached the decoder. Once they are captured and stored in the decoder, pages can be searched or browsed in much the same way that a videotext database can be examined. Thus teletext does permit active searching of transmitted information even though the system itself is one-way.

Access Delays. The information storage capacity of the decoder is the key to the speed of access and level of interactivity a teletext system can provide. Technically, decoders can be built to hold unlimited amounts of information, any page of which can be displayed instantaneously, much the same as a computer database. Practically, given the expense of computer memory and other components needed in decoders, their capabilities have to be compromised in order to keep the cost down.

Thus, in order to keep cost in the $200 to $300 range, typical decoders can retain only a few pages in memory for instant access, usually the menu, and must capture any others from the transmission cycle. Unless you select a page that happens to be in memory, the decoder must wait for the page to come by in the cycle. The amount of time the page takes to appear depends on where it is in the cycle. On a 100-page system, if you select page 100 just as page 1 is broadcast, the decoder will have to wait for 99 pages to go by before it can capture page 100 for you.

In a typical teletext system it takes from ten to twenty seconds to transmit a 100-page cycle, depending on how the system is designed. Thus, at most, you would have to wait about twenty seconds for your information. If you happen to request the next page being transmitted in the cycle, the wait is only about one-fifth of a second. Average wait for pages in a 100-page teletext magazine or database using low-price decoders is ten seconds. By adding decoder memory (and expense) and positioning pages carefully in the cycle, 200-page magazines have achieved comparable average delays.

Teletext in Europe

Origins in Britain. Like videotext, teletext was initially developed in Britain in the early 1970s. Engineers at the BBC and the Independent Broadcasting Authority (BBC's commercial counterpart) were researching television captioning systems to assist hearing-impaired viewers. Their efforts led to something more elaborate than simple captioning, and in 1974 field trials began of what they first called their video newspaper. Full-scale operations were underway two years later by the BBC and the IBA. BBC's version, called CEEFAX ("see facts"), was carried on two channels without advertising. IBA's system was dubbed ORACLE (Optional Reception of Announcements by Coded Line Electronics); like the IBA stations, ORACLE carried advertising. Neither system charges teletext user fees; CEEFAX is supported from regular BBC revenues (the annual TV tax paid by all set owners) and ORACLE's revenues come from advertising fees.

British teletext popularity grew rapidly in the early 1980s, and today ORACLE and CEEFAX together are the world's largest system, with nearly

3 million homes able to receive them. Important factors in this success are said to be the compatibility of the two systems, extensive market research done by the developers, the simplicity of the technology, and the ready availability of TV sets with inexpensive built-in teletext decoders.[12]

France and Elsewhere. In France teletext was launched in 1977, based on the same Antiope transmission standards used for French videotext. As is the case with Antiope videotext compared to Prestel videotext, Antiope teletext produces more detailed, higher-quality text and graphics than CEEFAX/ORACLE. It is also incompatible with the British system. French teletext at present reaches a relatively small number of mostly business users through national TV stations. Both users and information providers pay fees. Belgium, Italy, and Spain have also adopted the Antiope standard for state-run teletext systems under development.

Elsewhere, the exported version of the British teletext standard, marketed as World System Teletext, has become the de facto European standard. Eight countries, including West Germany with nearly 3 million terminals, have adopted the British system and developed programming paid for by government funding. All together, perhaps 5 million teletext-equipped TV sets could be found in European households in 1988, with most in Britain and West Germany. Content of most systems is similar to the British, with news and consumer information paramount, and most are free to use. As of 1986 only Switzerland had joined Britain in permitting advertiser-supported information.[13]

In Canada a teletext version of the Telidon system discussed earlier was tested several years ago. The Canadian Broadcasting Company ran teletext service trails in Toronto, Calgary, and Montreal in 1982 and 1983 in project IRIS (Information Relayed Instantly from the Source).[14] As with Canadian videotext, mass-scale systems have yet to be developed.

Teletext in the United States

Standards Problems. Teletext has yet to become a mass medium in the United States for reasons similar those that have already been mentioned regarding videotext. Lack of federal government involvement both in funding and research has left the field wide open and the research and marketing investment up to the private sector. After authorizing the service in 1983 the FCC declined, as with videotext, to adopt any existing system as the U.S. standard. Consequently several teletext experiments have been carried out by broadcast networks and other major media companies, some local TV stations, cable systems, and universities using incompatible systems.

The least expensive and also least graphically sophisticated is the World System Teletext (WST) standard developed in Britain and used in much of Western Europe. A more sophisticated standard has also been used—one based on both Canadian Telidon and French Antiope technology and compatible with both of them. Called the North American Broadcast Teletext Specification (NABTS), it was developed and backed by RCA, NBC, Time, and AT&T and used by them in trials. It is compatible with NAPLPS videotext and offers much more impressive graphics than WST. But NABTS terminals have cost as much as $1,000 in trial systems and usually work only with special TV monitors, not average home TV sets. WST terminals cost less than $300.

U.S. Systems. The longest continually operating U.S. teletext system is a noncommercial one: the captioning service begun in 1980 by the National Captioning Institute near Washington, D.C. This is not a full-screen color system and lacks user interactivity, but it is a true teletext system using the VBI for transmission of captions for the hearing-impaired viewer. Captions are inserted by NCI staff into videotaped or live programs and then decoded and displayed on the viewer's TV set. In 1988 about 250 hours per week of programs were captioned at the Institute, including all the broadcast networks and many cable services. The project is supported by government and private funds. Nearly 200,000 of the $225 decoders were in place in U.S. homes in 1988.[15]

U.S. commercial, mass-market teletext trials began shortly after CEEFAX/ORACLE debuted in Britain. In 1979 television station KSL-TV in Salt Lake City, Utah, began experimenting with their version of the WST system after seeing CEEFAX/ORACLE in Britain. KSL later switched over to NABTS-standard teletext and after years of experimentation now maintains one of the nation's few fully staffed local teletext services including news and local advertising. KSL's transmission consists of 100 pages of locally generated text and about 100 pages of national information received in the VBI of the network programming it receives from CBS.

CBS was involved in teletext trials of various types as early as 1979 and began inserting news, sports, and other national content into the VBI of its network feed in 1983. Although all CBS affiliates could receive Extravision, only four conducted local trials. The first to launch a local teletext trial featuring both Extravision and locally produced information and advertising was WBTV, Charlotte, N.C., which unveiled a 100-page local service in 1984. However, in Charlotte, as was the case in other NABTS-equipped trials, lack of affordable decoders severely limited the number of viewers who could use the service in their homes.

WBTV shut down the local portion of its service in 1986, leaving KSL, Salt Lake City, the only affiliate still transmitting local-plus-Extravision teletext in 1988.

Other major but short-lived U.S. teletext efforts in the 1980s included NBC's transmission of a fifty-page service similar to Extravision from 1983 to 1985; Time's experiments with full-channel teletext on cable systems in San Diego and Orlando;[16] local broadcast of a WST-type service (Electra) by WKRC, Cincinnati, Ohio; and another WST service (Keyfax), which was broadcast locally over WFLD-TV, Chicago, and nationally transmitted in the VBI of cable superstation WTBS, Atlanta. With the exception of Extravision and the NCI's captioning service, all these trials and numerous others were cancelled within a couple of years.

Decoder Dilemma. The reason most often cited in the trade press for teletext failures was the lack of affordably priced decoders. It was, and still is, a chicken-and-egg problem: without a national audience, no manufacturer will commit to mass production of decoders; without inexpensive, readily available decoders, no audience can be attracted. Teletext optimists continue to argue that it will take off as a national medium whenever inexpensive decoders are available.

Perhaps the enthusiasm for teletext persists at places like CBS and KSL because of teletext's two clear virtues as an electronic publishing medium. It is a relatively cheap technology for an operator to install, and it makes use of communication channels already in place: the VBI of a local broadcast or cable TV channel.

An Obstinate Audience

Perhaps the greatest challenge confronting developers of mass-market commercial videotext and teletext is developing services that will catch the imagination of a mass audience and ensure the financial support of advertisers. This will have to be done despite the limited interactivity of teletext and the presence of alternative forms of electronic publishing like the compact disc. In the United States achieving mass acceptance will mean changing the way the average person uses television at home.

This will require overcoming the prevailing attitude that television is watched for entertainment, not for information; the aversion of many people to reading text on a television screen more than moments at a time; and the convenience, portability, and low cost of newspapers and other print media.[17] If these obstacles to mass electronic text services are not overcome at least partially, videotext and teletext will not evolve

beyond specialized niche media serving only persons with specialized information needs.

Notes

1. These categories and the comparisons discussed below are drawn in part from Michael Rice, "Toward Enhancing the Social Benefits of Electronic Publishing: Report of an Aspen Institute Planning Meeting" (New York: Aspen Institute, 1987).
2. "Datapro Reports on International Telecommunications: The United Kingdom," Report #IT10-020-800 (Delran, N.J.: Datapro Research, 1988), p. 810.
3. See "Punching Up Wine and Foie-Gras," *Newsweek* (1 December 1986): 65.
4. John Tydeman and Ellen Jakes Kelm, *New Media in Europe: Satellites, Cable, VCRs and Videotex* (London: McGraw-Hill, 1986), p. 227.
5. Eugene Dimaira, "NT Readies Videotex Terminal," *MIS Week* (6 June 1988): 1.
6. Gary Arlen, "Videotex: People Still Don't Know Just What They're Missing," *Channels* (December 1988): 123.
7. See Jerome Aumente, *New Electronic Pathways: Videotex, Teletext and Online Databases* (Beverly Hills, Calif.: Sage, 1987), p. 69.
8. Steve Behrens, "IBM and Sears Redo Videotex as a Marketing Vehicle," *Channels* (March 1988): 15.
9. See "NT Readies," and Mark Rockwell, "U.S. West, Minitel Work on Gateway," *MIS Week* (5 May 1988): 15.
10. See Chapter 13 for a discussion of the constraints placed on the regional Bell operating companies as part of the deregulation of the telephone industry.
11. Tom Kerver, "Born Again?," *Channels* (1 May 1987): 58.
12. See Hans J. Kleinsteuber, et al., eds., *Electronic Media and Politics in Western Europe* (New York: Campus Verlag, 1986), p. 124.
13. See Tydeman and Kelm, *New Media,* pp. 211–20, for an excellent discussion of European teletext, on which this overview is partly based.
14. Aumente, *New Electronic Pathways,* p. 41.
15. Call the NCI at 1-800-533-WORD for information on how to purchase a decoder.
16. In full-channel teletext all 525 lines of the TV signal are used for text transmission, not just the VBI. Time's service offered 4,500 pages, transmitted as a separate TV channel on local cable systems.
17. For further discussion of these obstacles see Efrem Sigel, ed., *Videotext: The Coming Revolution in Home/Office Information Retrieval* (New York: Harmony Books, 1980), p. 131.

15

Transactional Services: Video Retailing

Since the first person used the newly invented telephone to call a local store to have groceries or other goods delivered over 100 years ago, telecommunications technology has played an important role in retailing. Until recently that role was fundamentally unchanged. Now advances in telecommunication and computer technology have spawned a wide range of electronic shop-at-home experiments by innovative entrepreneurs and powerful national merchandisers. Combining electronic media with mass marketing expertise in new ways, these efforts may ultimately lead to significant changes in the way millions of consumers in the United States and elsewhere select and purchase goods. In this chapter we survey some current teleshopping efforts.

Varieties of Electronic Retailing

Americans have been shopping at home for about a century, ever since Sears and Montgomery Ward published their first mail-order catalogs. Direct-mail retailing has become a $50 billion industry, one characterized until recently by several thousand companies that did business by mail only. The growing popularity of shopping at home has led national retail chains and department stores into the direct mail business as well, partly to protect their $150 billion annual in-store sales from further encroachment from catalog companies. The popularity of catalog shopping and other shifts in consumer buying habits coupled with developments in communication technology have laid the groundwork for the present interest in electronic shopping by major retailers. Direct sales to customers, free of the overhead involved in maintaining retail outlets, is also a primary attraction of electronic retailing. Broadly speaking, electronic retailing today takes two major forms.

Video Shopping Channels

The first approach, which has gained considerable notoriety since the mid-1980s, is the video shopping channel. In this case, the merchandiser becomes a television programmer, producing what amounts to a continuous stream of commercials describing individual items for sale to consumers who happen to be watching. The shopping programs are broadcast over local TV stations, usually low-rated UHFs, and distributed nationally by cable.

The concept of mass-marketing inexpensive merchandise over television is not a new one; anyone who recalls the Vegematic or Popeil Pocket Fisherman of late-night television fame is familiar with the core concept of a video shopping channel. Today's continuous selling may be a bit smoother and have certain talk-show amenities, but the basic idea is still the same. Although the selling on shopping channels may be organized around certain categories of goods or speciality items, typically the presentation of merchandise is random. This is because the items the video retailer has accumulated for sale are most often purchased as large lots of overstocked items or other distressed goods. Stereo speakers might be followed by fishing reels, which in turn are followed by kitchen cutlery or costume jewelry.

Shopping channel presentation of merchandise can range in sophistication from still pictures accompanied by an announcer's voice to a talk show format involving celebrity hosts presenting items modeled or demonstrated. Program formats can vary as well, including games, contests, and other promotions designed to hold viewers' attention. Once a decision to buy is made, the shopper places the order by phone. The order is handled most often by a customer service representative, but increasingly, automated systems are being put to use. In this case a computer "asks" for the necessary information, which is keyed in using the shopper's touch-tone telephone. In both cases orders are charged to credit cards.

Video shopping channels reach their audience using two major outlets: over-the-air broadcasting stations and cable systems. On cable, the shopping channels are normally offered as part of the basic program package, either as a twenty-four-hour a day service with its own channel or as filler material on a channel featuring other kinds of programming. Cable operators receive a commission on all purchases made by their subscribers. On some cable systems shopping channel commissions can add up to significant income. According to one trade magazine some systems receive as much as $17 to $20 per year per subscriber from certain popular video shopping services, and the industry as a whole took in $100 million in commissions between 1985 and 1988.[1] In some major television markets, independent UHF stations have been purchased by teleretailers and converted from their usual fare to shopping channels, carrying the same service that may also be distributed in other areas via cable. The leader in this tactic has been the Home Shopping Network, which has purchased fourteen UHF stations to be used as full-time shopping channel outlets.

Services in the United States and Europe. Although some two-way cable systems experimented with local and regional TV shopping beginning in the late 1970s, the video shopping channel boom began with the overnight success of the Home Shopping Network when it first began national satellite distribution in 1985. By early 1988 HSN reportedly had sold over $1 billion worth of merchandise and its success had spawned a half-dozen major competitors plus numerous other ventures. Together with HSN, video shopping channel sales were expected to approach $2 billion in 1988. Put in perspective, video shopping revenues skyrocketed in just three years from negligible to more than the total advertising revenue of the cable industry (about $1.2 billion in 1987). HSN dominated revenue, generating about half the sales itself.

Several of the major competitors joining HSN in the rush to cash in on video shopping's success have been funded by cable companies, enticed by the potential revenues from sales commissions. The Cable Value Network (CVN) and the QVC Network are both funded by a major multiple cable system operator, TCI. The Fashion Channel and the Travel Channel also are underwritten by cable companies. Although HSN leads dramatically in sales, it is only in third place in number of homes reached (15 million); CVN, Consumer Discount Network, and Video Mall Network each reached about 20 million households in 1988, a little less than a quarter of the U.S. TV audience.[2]

In Europe video home shopping debuted in 1987. Backed in part by two major British retail chains, the "Space Shopping" service is transmitted fifteen minutes Monday through Friday on the pan-European cable network Sky Channel, which reaches around 9 million homes in sixteen countries. The talk-show format of "Space Shopping" and its array of kitchen gadgets, small appliances, jewelry, and sports equipment is similar to U.S. programs. However, European home shopping services face difficulties never imagined by U.S. shows. The regulatory, economic, and cultural difficulties that arose when selling simultaneously in so many countries were referred to as a "hornet's nest" by the channel's director, requiring a great deal of effort to overcome.[3] For example, the program's length is limited to fifteen minutes because of rules in several countries limiting the amount of advertising TV channels—even cable—can carry per hour.

Laws regulating French broadcasting were envoked to protest the debut of Europe's second video shopping service on the newly privatized TF-1 national television network. A thirty-minute daily shopping program, "Magazine de L'Object," was launched in fall 1987, similar in style and content to "Space Shopping" and its U.S. counterparts. Critics objected that French law was not clear as to the status of shopping programs and that they should be forestalled while consumer rights issues were examined. Late in 1987 other shopping channels were also planned in France.[4]

Audience Limitations. Although it obviously requires an active involvement on the consumer's part to pick up a phone and order merchandise, a video shopping channel is essentially a passive experience from a marketing/sales point of view. Not knowing which items are coming up next, the consumer sits in front of the TV watching a more or less random presentation of merchandise until something strikes his or her fancy. Consequently, research indicates that impulse buying dominates video shopping channel sales, with the great majority of purchasers characterized as "bargain hunters."[5] Furthermore, these bargain hunters have made video shopping services a $2 billion industry, but they represent only a small portion of the potential buyers. Only 15 percent of the nation's cable subscribers bought anything in 1987.[6]

Leading services like HSN are making an effort to expand their audiences beyond this small core of bargain-hunting regular viewers, to a broader audience that will consider video shopping a legitimate substitue for a trip to the mall. Efforts concentrate on subscriber awareness and program content. According to one cable company's 1987 research, 80 percent of nonbuyers had never watched a video shopping channel and 50 percent had not even heard of one.[7] Consequently, video shopping channels are purchasing commercial time on other channels and using other advertising media to increase awareness of their services.

For example, HSN budgeted $36 million for marketing in 1987.[8] It also is finding ways both to improve its product mix by cutting deals with name-brand retailers to market their merchandise and also to move away from the fast-talking, horn-tooting "cubic zirconium" huckster image. Some services are also experimenting with segmenting product presentation—that is, selling certain types of merchandise at specific times of day, so viewers seeking a particular item will know when to watch.

Videotext Shopping Services

Most shoppers have seen computers used in retail stores to record purchases and place orders for merchandise. Videotext shopping services essentially extend this in-store use of the computer to the customer's home. Given a videotext terminal or personal computer, appropriate software, and a user-friendly system for browsing and ordering from a store's stock, customers can help themselves. Whereas the usual computerized inventory is merely a coded list of stock on hand, a computer shopping service takes this list and embellishes it with color, illustrative graphics, and extended product descriptions to encourage purchase.

Today, most videotext shopping services are available through a third party, not the retailer directly. Typically, videotext services such as those mentioned in Chapter 14 carry shopping services from a number of retailers as part of their total package of databases. The subscriber uses a videotext terminal or personal computer to access the service via phone lines. A two-way cable system can also provide such services on designated channels by acting as a gateway or link between cable subscribers and shopping services. On two-way cable systems two-way converters readily take the place of personal computers as shopping terminals. Hybrid systems are even being tested for one-way cable systems, in which the upstream data link between customer and cable headend is provided by touch-tone phones.

Once the shopping service is called up on the computer monitor or television receiver screen, the buyer is presented a menu that invites selection of major categories of merchandise, ultimately leading through a series of such menus to the particular item desired. Products are ordered according to available instructions and charged to store accounts or bank credit cards. Even delivery dates and times can be arranged by computer.

Greater Interactivity and Efficiency. Compared to video shopping channels, using a videotext shopping service is very much an interactive experience: no merchandise can even be viewed until the consumer chooses a category. Videotext home shopping presumably encourages browsing—electronically scanning the pages of the latest mail-order catalogue. Unlike a video shopping channel, videotext shopping allows a rapid, focused search for a specific type of merchandise or particular item. This is a major advantage in efficiency compared to video shopping channels and is thought to be a key to broadening the appeal of home shopping beyond impulse buyers and bargain hunters. Furthermore, while time is a major constraint on the amount of merchandise that can be presented on a video shopping channel, on a videotext shopping service the amount of merchandise displayed is limited only by the storage capacity of the computer—virtually limitless—and the expense of designing and inserting pages for each item into the database.

Systems in Operation. Most large-scale videotext services in Britain, France, and the United States have some transactional capability that allows merchandise to be selected, ordered, and paid for. France's Teletel videotext system has introduced the use of "smart cards" to take this capability one step further. A smart card is similar to a credit card, except it contains a microchip instead of the familiar strip of magnetic tape. The chip is programmed with a variety of user financial information so the smart card is able to keep track of how much money it has spent, where it was spent, and how much remains. Once depleted, it can be financially "refilled" by the user's bank's computer via a Minitel terminal. With their smart cards, Minitel users purchase everything from wine and groceries to clothes and concert tickets.

In the United States the major videotext system, CompuServe, offers videotext shopping through its Electronic Mall feature. Computer owners subscribing to CompuServe can use credit card numbers to order merchandise from retailers and catalog companies. The Electronic Mall appears similar to the transactional features offered for a short time on Gateway in Los Angeles and Viewtron in Miami, the biggest videotext trials in the United States during the 1980s. Although both systems were shut down in less than two years, they did demonstrate that shopping at home appealed to a majority of subscribers and pioneered many technical aspects of product presentation on the video screen (see Chapter 14). Despite these efforts, relatively little merchandise was being sold via videotext shopping services in the United States in 1988.

Prodigy. Two major efforts were underway, however, to revitalize the videotext shopping industry. The first is aimed at home computer owners exclusively (IBM, Apple, and compatibles) and is a joint venture of Sears and IBM, called Trintex. The Trintex videotext system, called Prodigy, was scheduled for test marketing in Hartford, Atlanta, and San Francisco in mid-1988. Prodigy is technically similar to CompuServe and other information services. However, in addition to the usual sorts of consumer-oriented news, sports, and other information, Prodigy emphasizes transactional services more than its predecessors and allows subscribers to purchase merchandise from some seventy major retailers and brands. These and others will also buy advertising space on information pages. Trintex hopes to convince the majority of computer owners who do not subscribe to videotext services that Prodigy is economical and easy to use. To that end it has made arrangements with a major manufacturer to supply the modems needed to connect the subscriber's computer to phone lines and will charge only a flat monthly fee ($10 to $15) for Prodigy use, avoiding hourly line charges.[9]

Telaction. The second major videotext shopping venture takes an entirely different approach. Telaction, being introduced in the Chicago area in 1988, is the product of several years' research and investment by J.C. Penney. Rather than home computer owners, Telaction is aimed at a much larger audience, the 50 percent of U.S. households subscribing to cable.

Telaction uses the cable system to link cable subscribers with its operating center near Chicago in Schaumberg, Illinois, where pictures of merchandise and ordering information are stored on video disc (see Chapter 7). Because the video disc can store sound as well as pictures, Telaction offers a unique capability to retailers: audio can accompany the video presentation when a user selects a particular page. A person interested in buying a wrist watch can hear it described as well as see it; a prospective record buyer can hear a short sample of the music.

Cable subscribers select the Telaction channel on their TV sets and call a local telephone number appearing on the screen. The Penney catalog and product information of three dozen other retailers and service companies are transmitted to the cable subscriber's home via the cable channel. The user interacts with these databases using the telephone keys, following instructions on the usual sorts of menus to select and order merchandise.

Telaction does not require two-way cable technology or special equipment in the subscriber's home—just standard cable service and a touch-tone telephone. It does, however, require installation of specialized equipment at the cable system headend and frame storage devices on the cable system lines every block or so that to hold and supply the video images selected by the subscribers.

In-Store Videotext. A third approach that may help to revitalize videotext shopping is the use of public-access kiosks or booths in malls and hotels. As Chapter 14 mentioned, videotext kiosks providing shoppers with directions and travelers with automated check-in/check-out services are appearing in major metropolitan areas across the United States. Kiosks are also being installed in stores to provide product demonstrations and automated purchasing from catalogs. Grocery stores are increasingly installing kiosks sponsored by food companies that provide cooking tips and other product information.

As kiosks provide more transactional services, according to one marketing research firm, they will familiarize the public with videotext as a shopping medium. This will provide the impetus needed to help in-home services like Prodigy and Telaction become national mass media, used in average households to make their normal shopping less time-consuming.[10]

Electronic Shopping in the 1990s

Just as automatic teller banking slowly gained acceptance and became a common convenience, some form of electronic shopping can be expected to gradually find a niche to fill among consumers. One analyst claims that public access and in-store kiosks alone will constitute a billion dollar industry as early as 1990.[11] As mentioned earlier, video shopping channels are already a $2 billion industry.

However, observers predict that the ten or so competing services will soon be reduced to a much smaller number after a major shake-out, as the business passes from the take-off stage to a more mature phase. Although video shopping channels are expected to remain in some form, the number of major services and their significance remains difficult to predict. It is likely, however, that the shake-out survivors will make an effort to broaden their appeal beyond the small market segment of impulse buyers now comprising their core audience.

Perhaps the biggest question mark should be reserved for the videotext shopping services. Minitel appears to have established itself in a monopoly regulatory environment in France, but the U.S. picture is quite different. Despite the expenditure of several hundred million dollars by major media companies, retailers, and other communication companies, videotext in general and videotext shopping in particular have yet to make an impact on the information/shopping activities of the typical American consumer.

Videotext promoters have been compared to "fans of a baseball team that dwells in the cellar; to the believers there's always next year."[12] The search for the right combination of services, information, ease of use, and price may well continue into the 1990s. Particularly interesting to watch will be the early going of services like Telaction and Prodigy in the United States and the continued development of Minitel in France.

Notes

1. Gary Arlen, "Getting All That Gold Makes It Easier to Love All That Glitz," *Cablevision* (9 May 1988): 68.
2. Cecilia Capuzzi, "A Year after the Frenzy, Order Starts to Set In," *Channels* (December 1987): 11.
3. Laurel Wentz, "Home Shopping Arrives in European Markets," *Electronic Media* (11 May 1987): 44.
4. Lisa Bannon, "French Home Shopping Debuts; Consumer Group Seeks Rules," *Multichannel News* (12 October 1987): 22.
5. Ellis Simon, "Home Shopping Faces Maturity," *Cable Marketing* (January 1988): 28.
6. "Ops Find Only 15 Percent of Cable Subs Shop at Home," *Multichannel News* (24 August 1987): 38.
7. Ibid.
8. "Home Shopping Faces Maturity," 29.
9. See Steve Behrens, "IBM and Sears Redo Videotex as a Marketing Vehicle," *Channels* (March 1988): 15.
10. Tom Kerver, "Born Again?," *Cable Television Business* (1 May 1987): 58.
11. Ibid.
12. Gary Arlen, "Videotex: People Still Don't Know What They're Missing," *Channels* (December 1987): 123.

PART VI

The Telecommunication/ Transportation Tradeoff

16
Teleconferencing

Half the airline tickets sold in the United States each year are used for travel to business meetings.[1] Transporting, housing, and feeding traveling executives or other employees costs some major corporations as much as $100 million per year.[2] All together, business meetings and conferences in the United States alone are said to cost perhaps $250 billion per year.[3] Furthermore, these direct costs do not begin to reflect the hidden cost of workhours lost while en route from one meeting to another or of lowered productivity from travel fatigue. This chapter surveys the increasing use of teleconferencing to reduce these travel expenses and the array of technologies and services that makes electronic meetings possible.

Developing Interest in Teleconferencing

Beginning with the invention of the telephone more than a hundred years ago, the notion has persisted that electronic communication can take the place of much business travel. However, even after improved switching technology made possible conversations among persons at more than two locations, the phone by itself never became a panacea for the problem of business travel; it lacked the visual dimension so important in most forms of human interaction. Beginning in the 1930s, when practical electronic video technology became available, government and educational agencies experimented with early forms of video teleconferencing, demonstrating its technical feasibility. Even so, in the years that followed videoconferencing remained mostly a curiosity because expensive equipment and the requirement for special high-cost transmission links made it impractical for even major corporations.

At the 1964 New York World's Fair AT&T's Bell Laboratories introduced a major technological breakthrough, the Picturephone, a compact TV-telephone console that could transmit live video and audio for the first time over standard telephone lines. Unfortunately, because its low-resolution black-and-white picture did not provide the detail needed for document transmission, the Picturephone failed to catch on commercially.

Finally, in the 1970s several trends converged that led to widespread interest and subsequent investment in many forms of teleconferencing. Domestic communication satellites became available, providing an economical long-distance transmission link for audio and video that did not require the use of expensive terrestrial telephone circuits. Communication equipment—particularly satellite receiving and transmission dishes, telephone terminals, and video cameras—declined in price while simultaneously improving in capabilities. Technical breakthroughs in transmission technology including digital and fiber optic systems made new capacity available on phone circuits. FCC and other government agency actions led to the deregulation of the telephone industry, allowing opportunities for new manufacturers and service providers to offer a variety of new business communication services in competition with existing companies. Lastly, the cost of transportation began a dramatic rise in the mid-1970s and the cost of travel became a significant problem for many large companies and government agencies. Coupled with a general concern about declining productivity and competitiveness in American industry, these trends provided the impetus for the growth of today's teleconferencing industry.

By 1985 U.S. companies were spending over $100 million a year to stage many types of electronic meetings and purchase teleconferencing equipment.[4] National videoconferences involving dozens of cities and tens of thousands of participants are common today, and international—even worldwide—meetings on both small and grand scales are increasingly being attempted.

Hotel chains like Holiday Inn and Hilton have invested substantially in teleconferencing facilities and offer networking services to attract and hold corporate clients. Many universities that have been long-time pioneers in the use of television for one-way teleclasses off campus are using teleconferencing technology to provide two-way communication between instructors and students and to link campuses, corporations, and research agencies in video networks. Many major corporations have found the travel savings from electronic meetings to be substantial enough to justify their purchase of permanent videoconferencing facilities. Hundreds of others regularly engage outside firms to set up special teleconferences for important company meetings and corporationwide announcements such as new product introductions. Although it is clear that electronic meetings are not appropriate in every situation, teleconferencing in its many varieties is rapidly becoming an accepted form of business communication.

Four Types of Teleconferencing

Teleconferencing includes electronic meetings as simple as a three-way telephone conversation or as complex as a multinational videoconference. A wide variety of equipment and applications exist, but they generally can be thought of as falling into four basic categories of increasing expense and technical complexity.

Audio-Only Teleconferencing. Audio-only teleconferencing is a familiar business communication tool used by about half the country's organizations on a regular basis.[5] All that is needed to set up an audio-only teleconference between two locations is a telephone at each site. Multiple participants at each site use extension phones or gather in the same room using a microphone and speaker, which are built-in or connected to the telephone receiver. Voice-activated individual microphones and other auxiliary audio equipment are added on higher-cost permanent installations.

Adding a third or more sites to an audio teleconference requires the use of a bridge, an electronic device that links the participating telephones into a single network, allowing each participant to speak to, and hear, all the others. In addition to providing the hook-up, the bridge amplifies and balances the input and output from each location and filters out static or echoes that might be repeated throughout the network.

Conference telephones with built-in bridges are available that will link a limited number of sites and switchboards (called PABXs) can be purchased with bridging capability for larger conferences. As an alternative to investing in equipment, teleconference calls can also be arranged through local telephone companies (dial-up conference service) or independent conferencing services, which interconnect multiple sites with their own switching and bridging equipment after an order is placed in advance.

Teleconferencing with Augmented Audio. For certain kinds of meetings words alone are not adequate. In visually oriented or technical fields it is often necessary to have available diagrams, blueprints, tables, sketches, photos, or documents. Devices that can transmit replications of such graphic material over phone lines are referred to as audiographic equipment. The oldest and most familiar audiographic device is the facsimile machine or fax.

A basic fax system consists of two devices somewhat resembling a small photocopier or computer printer, one at each conference site. A document, drawing, or photo inserted into one machine is scanned and reproduced on paper by the other machine, providing participants a hard copy of material being discussed or presented. Material can be sent in either direction during the teleconference. Fax machines are inexpensive to use because they are linked by ordinary phone lines. However, because of the limited data capacity of such lines, document transmission can take from one to five or six minutes per page. Because transmission delays limit the use of fax machines during conference calls, documents are sent before meetings begin when possible.

A second group of devices, called telewriters, allows conferees to share sketches or notes in real time, as they are being created. An electronic tablet and television monitor at each site are linked by ordinary phone lines. A special electronic stylus, perhaps a light pen, is used to draw on the tablet or directly on the television screen. Drawings or figures made in one location are seen simultaneously at the other. Multiple color images can be produced and participants at each conference location can, with a little practice, modify or erase portions of the same picture.

A third way to add visual materials to audio teleconferences is the use of computers. By interconnecting personal computers or using terminals attached to the same central computer, conferees can create or call up from computer memory any graphics or printed material their system retains and displays. Materials can be discussed, created, or modified in real time by several participants whose phones and computers are interconnected.

Slow-Scan Videoconferencing. The term *videoconferencing* refers to the transmission of images from one site to another using video cameras, rather than computers, telewriters, or fax machines. Two kinds of videoconferencing are possible, each with different uses and advantages: in full-motion videoconferencing, live images are transmitted in real time; slow-scan video (also called freeze-frame, narrow-band, or captured-frame video) is the transmission of single video images, one frame at a time.

Slow-scan uses inexpensive, ordinary telephone circuits, and the equipment itself usually takes the form of a self-contained table-top or roll-around unit that can be plugged into a conventional phone outlet. Compact, low-resolution monochrome units no larger than a toaster are even available for home use.

A slow-scan system allows transmission and discussion of any image the camera might be aimed at. Depending on the technical sophistication of the system and the resolution desired, single frames take from ten seconds to a minute to transmit. The camera is focused on the desired subject, perhaps a photograph, drawing, or chart, a button is pushed, and the video snapshot is taken. This image is converted into a signal that is transmitted line by line to receivers at the other conference sites.

At the receiving end each line of the image is painted on the screen top to bottom as it is received or captured by frame-store circuitry and displayed after it is completed to avoid this waterfall effect. The image can be changed only by taking another video snapshot; there is no way to modify an image once it has been transmitted. Although still pictures of participants are possible, they are usually limited to group shots or carefully selected shots of main speakers. Random shots of conferees can result in embarrassing images of people in midyawn or with their eyes closed. Slow-scan video technology is usually used for document, photo, or figure transmission. If high resolution is needed for certain types of printed materials, a facsimile machine may be a practical alternative.

Full-Motion Videoconferencing. Transmission of a live television picture requires that a tremendous amount of data be moved at very high speed: thirty frames of video per second. While conventional microwave and satellite channels can handle a normal video signal, telephone lines cannot. If the sender wants to use existing data transmission lines for full-motion videoconferencing, the signal must be specially processed before transmission.

A device called a codec (coder/decoder) is used to transform the TV signal from its normal form into digital data and then compress it. Compression electronically shrinks or compacts the video signal. This allows full-motion video transmission over high-capacity data lines, the kind that may already be in use for computer communications by a company wishing to set up a videoconferencing system. Codecs also provide the advantage of encrypting or scrambling the transmission, to avoid eavesdropping on videoconferences and compromise of proprietary information.

Although compression technology was improving dramatically in the late 1980s, full-motion video still could not be transmitted over inexpensive voice-grade telephone circuits, as could be done with slow-scan video. The video production and signal transmission equipment required for full-motion video make it many times more expensive than slow-scan.

Today there are two primary applications of full-motion video in teleconferencing. By far the most common is one-way video transmission accompanied by two-way audio. This arrangement is often used to transmit a speech, lecture, or product presentation from a single location to multiple receiving sites. It can also be used to bring a prominent speaker into a large meeting at a single location, such as the president addressing a trade convention or an evangelist preaching to a stadium full of people.

The basic arrangement at the sites is the same regardless of how many receiving locations there may be. The originating site is equipped with one or more video cameras, monitors for observing the outgoing picture, plus switching, lighting, graphics, and audio equipment similar to that needed by a commercial TV station for remote or studio transmission of an event or interview. Receive sites each have monitors or large-screen projection equipment and usually are equipped with microphones allowing communication upstream to the presenter via ordinary telephone circuits.

In two-way full-motion videoconferencing all sites are equipped for video transmit and receive capability, so pictures are simultaneously being transmitted in both directions at all times. The need for every site in effect to operate as a fully equipped remote TV studio makes two-way the most expensive and, consequently, the least widely used type of full-motion videoconferencing. Permanent videoconference rooms are usually designed to look less like a TV studio and more like a traditional meeting room. Ideally, monitors, cameras, lights, and cables are built into the walls surrounding a desk or conference table and light levels are kept as near as practical to normal to make the technology as unobtrusive as possible. Equipment cost for a single videoconference facility ranged in 1988 from $50,000 to $150,000. Furnishings, renovations, and construction add substantially to this hardware expense. On the other hand, a video receive-only site can be installed for well under $10,000.

Consequently, though the number of corporate branch offices, dealerships, and even retail outlets with full-motion receiving capability reportedly numbers in the thousands and is still growing, the number of large companies with permanent two-way videoconferencing facilities probably is only a few hundred. Those that have built extensive multi-site two-way networks probably number only in the dozens.

U.S. Teleconferencing: Commercial and In-House

The less expensive or complex forms of teleconferencing, like audio-only, audio-graphic, and slow-scan, are usually accomplished with in-house (user-owned and -operated) equipment, but this is not the case for full-motion videoconferencing. With permanent facilities uneconomical for most companies, a service industry has sprung up, and most full-motion videoconferencing is done using third-party services. In 1980 Holiday Inn was the first major hotel chain to offer nationwide videoconferencing services and now has sites in more than three hundred hotels and motels. They have since been joined by Sheraton, Marriott, Hilton, and other business-oriented chains. AT&T has offered its Picturephone Meeting Service since 1982 at facilities in several major cities.

For very large groups or other situations in which the usual videoconference room would not be suitable, service providers will come to you. A pioneer in this field has been the Robert Wold Company, which leases large amounts of satellite time and resells it to individual clients on an as-needed basis. Wold also provides video equipment and production personnel who arrange and produce one- and two-way video-conferences on virtually any scale. The Wold Company and other commercial videoconference services make it possible for nearly any business or trade group to hold electronic meetings without equipment investment or specialized staff.

European Videoconferencing

In Europe, as in the United States, various forms of audio teleconferencing are widely in use, while videoconferencing, though growing, is still in relatively early stages of development. Aside from a few major corporations with in-house facilities, videoconferencing is available mainly through commercial or PTT-owned service companies.[6]

Reportedly the oldest videoconferencing service in continuous operation is Britain's Confravision system. First offered in 1972, Confravision is a service of British Telecom, which has established studios in several major cities and some mobile facilities as well. An enhanced version of this service, called VideoStream, began in 1985 and offers international links between Britain, the United States, and several major cities on the continent.

In France, Visio-Conference interconnects more than two dozen cities and is operated by the national PTT. As France completes its long-range plan for wiring a significant portion of the nation's homes with a high-capacity fiber optic cable system, videoconferencing may become a common household service, mirroring the popularity of Minitel (discussed in Chapter 14). Commercial videoconferencing studios are also allowed to interface with the state-operated Visio-Conference service.

Sweden's system, Teleconference, provides two-way videoconferencing between most large cities in that country and international connections via satellite. Elsewhere, videoconferencing systems are in earlier stages of development, with national or commercial services (or both) planned in most European Economic Community member countries. In some cases PTT-funded videoconference networks await the installation of upgraded communication circuits; in others, commercial networks have been thwarted by regulations prohibiting competition with state telecommunication monopolies.

Finally, as is the case in Europe with other new technologies discussed in this book, incompatible equipment standards contribute to delays. Even when countries share the same video standards the use of incompatible codecs has limited the ability to establish international videoconferencing networks.

Electronic Meetings in the 1990s and Beyond

After decades of experimentation, it appears that teleconferencing has passed over the line from novelty to business communication tool. As equipment and transmission costs continue to decrease and private-sector and government organizations seek ways to reduce transportation costs and improve productivity, the long-term prospects for the growth of the many forms of teleconferencing seem excellent. As Table 16-1 indicates, it is also expected that videoconferencing will play an increasingly larger role as users upgrade from audio-only to video systems.

As is often the case with new electronic media, adoption of teleconferencing is being driven by the obvious immediate benefits. Research generally shows that the most cited reason for electronic meetings is savings in travel expense and time.[7] However, not yet clear are the longer-term impacts of electronic meetings on the way we communicate in the workplace and the larger implications of this for the nature and organization of work activities in future society.

It is already apparent however, that along with the benefits some counterbalancing effects must be considered when planning the adoption of electronic meetings in place of traditional business travel and

Table 16-1. Changing Teleconferencing Expenditures

Type of System	*1980 ($50 million)*	*1985 ($250 million)*	*1990 ($900 million)*
Audio	65%	49%	15%
Audio-graphic	20%	30%	35%
Video	10%	15%	40%
Computer	5%	6%	10%

Source: *Communications News.*

face-to-face communication. Researchers have pointed out, for instance, that while electronic meetings can be set up on shorter notice than face-to-face meetings, this instant availability can lead to sloppiness in planning meetings, including who actually needs to attend and what the agenda should be.[8]

Research indicates that electronic meetings can expand the breadth of communication in an organization, providing broader input for decisionmaking and giving members a sense of familiarity with their distant counterparts and a feeling of management participation. But if not carefully managed, broader participation can also bring to the surface underlying organizational conflicts, aggravating differences and conflicts between units and personnel. On the plus side, electronic meetings tend to be more focused and task-oriented and usually shorter than in-person sessions. But there is a fine line between an orderly meeting and a repressive one. Finally, while teleconferencing clearly reduces travel expenses, it obviously increases communication expenses.[9] To a large extent the continued growth of teleconferencing depends on managers' ability to balance these benefits, drawbacks, and costs.

Notes

1. Mia Amato, "Videoconferencing: A Flashy Technology That Pinches Penney's Pennies," *Channels* (December 1986): 67.
2. James Martin, *Telematic Society: A Challenge for Tomorrow* (Englewood Cliffs, N.J.: Prentice-Hall, 1981), p. 161.
3. Amato, "Videoconferencing."
4. "More Firms Rely on Video Linkups," *Los Angeles Times,* 11 March 1985, sec. IV, p. 1.
5. Lynne L. Svenning and John E. Ruchinskas, "Organizational Teleconferencing," in Ronald E. Rice, *The New Media* (Beverly Hills, Calif.: Sage, 1984), p. 241.

6. For more information on European systems, see "Teleconferencing," Report #IT40-625-101 (Delran, N.J.: Datapro Research, April 1987).
7. See Frederick Williams, *Technology and Communication Behavior* (Belmont, Calif.: Wadsworth, 1987), p. 91.
8. Robert Johansen, Jaques Vallee, and Kathleen Spangler, *Electronic Meetings: Alternatives and Social Choices* (Reading, Mass.: Addison-Wesley, 1979), p. 131.
9. For a discussion of the expenses associated with teleconferencing, see Rob Drasin, "Ten Cost Factors for Teleconferencing," *Video User* (March 1983): 6.

17

Telecommuting

In the United States and other advanced industrialized nations today, many jobs involve primarily the creation, manipulation, storage, or retrieval of information. In the United States it has been estimated that about 50 percent of the workforce is information workers, with a significant portion of these workers traveling to their workplace primarily to sit at a computer terminal or personal computer. A number of pioneering organizations have shown that, given proper telecommunication technology and skilled management, many such employees can telecommute—that is, work at home linked electronically to the office. This chapter discusses the development of telecommuting and why its further growth presents more managerial and social challenges than technological ones.

How Interest in Telecommuting Developed

Since the early 1970s the cost of telecommunication and information technologies and services has steadily dropped, but economic and social trends have converged to increase interest in telecommuting. They include (1) the rising cost of commuting, (2) a changing workforce, and (3) corporate restructuring.

The Costs of Commuting. Perhaps the most dramatic development leading to interest in telecommuting was the energy crisis of the 1970s and the rising cost of fuel over the last twenty years. This has painfully increased public awareness of the cost of a production system requiring millions of workers to commute by car from their homes to centrally located workplaces.

As early as 1974 a benchmark study sponsored by the National Science Foundation examined these costs in detail, pointing out the energy-saving potential of telecommuting.[1] The study found that a sample of 2,000 insurance company employees in Los Angeles commuted an average of twenty-one miles a day, round trip, to work—a total of more than 12 million miles a year. In 1974 prices this represented nearly $2 million in gasoline and other transportation costs.

Although initially coming out of employees' pockets, these costs were eventually borne by their employers in the form of higher wages and inevitably passed on to consumers. A number of hidden costs of this system also were passed on, including the expense of building and maintaining employee parking facilities and the productivity lost by requiring the labor force to sit stalled in rush-hour traffic day after day. In the insurance company study the 2,000 employees together logged almost fifty years on the L.A. freeways each calendar year. Projecting the L.A. results to national scale, researchers concluded that foreign oil imports could be completely eliminated by converting only 1 to 14 percent of automobile commuters to telecommuters.

In the years since the L.A. study, average commutes in urban areas have gotten longer. Rising real estate prices have driven families further into the suburbs to find affordable housing. As traffic congestion increases and workers spend more time and energy parked in rush-hour traffic, practical ways to improve the present system are receiving increasing attention. They include decentralizing business operations to the suburbs, improved mass transit, more flexible commuting schedules, and telecommuting.

The Changing Workforce. Important demographic changes in the American population are underway and having an impact on the labor force, leading many employers to consider alternative or nontraditional working arrangements. For example, the number of two-worker families increased from 8.3 million to 13.4 million between 1976 and 1986, according to Census Bureau reports. Many of those two-worker families have small children. From 1976 to 1987 working mothers of young children increased from 36 percent to 50.8 percent of the workforce, with the percentage expected to continue rising in the coming decade.[2] The workforce is also aging and exhibiting less traditional attitudes toward work, including less organizational loyalty and less willingness to relocate or tolerate long commutes.

These demographic trends have also been accompanied by a growing shortage of certain white-collar skills, especially secretaries, typists, and computer programmers. Consequently, along with more flexible working hours, better daycare benefits, and shared jobs, many employers are experimenting with telecommuting in an effort to attract and hold nontraditional workers.

Corporate Restructuring. After an era of mergers and expansion during the 1960s and 1970s, many large U.S. firms retrenched in the 1980s, dropping unrelated lines of business and generally reducing overhead to improve productivity and competitiveness. In a quest for more flexible hiring and staffing, companies have converted many traditionally full-time positions to part-time or temporary slots. This shift toward a disposable workforce, while controversial, has also created a climate more open to innovations like telecommuting.

Telecommuting Today

Estimates of the number of telecommuters in the United States vary widely. In 1985 the Department of Labor estimated as many as 1.5 million workers were part- or full-time telecommuters out of more than 108 million total workers. The same year, the *Wall Street Journal* placed the number nearer to 100,000.[3] Estimates of the number of companies involved varied from 300 to 450.

Among the first companies in the United States to experiment with telecommuting have been insurance firms and computer-related companies, both of which hire large numbers of clerical and data-processing employees. Among major firms with experience in telecommuting dating from the early 1980s are Aetna Life & Casualty, Investors Diversified Services, Control Data, Digital Equipment, and Data General. Several regional Blue Cross/Blue Shield operations have telecommuting programs. Perhaps the largest commitment to telecommuting comes from a national retailer, J.C. Penney. Penney has fourteen regional telephone centers to take and process catalogue orders. About 200 customer service representatives handle customer inquiries and orders using terminals linked to the regional processing centers.[4]

Telecommuting plans vary considerably by company. Some hire telecommuters as full-time employees with all normal fringe benefits; others classify them as independent contractors, working part-time strictly on a wage basis without benefits. Some companies provide terminals to telecommuters; others require that they be rented or purchased. Telecommuter pay may be higher or lower than regular employees get for the same work, depending on these variables and others.

The firm with the longest ongoing telecommuting program is not American but British. A company that designs computer software and related products, F International, was established in 1962 by a woman who quit her regular job as a programmer to raise a family. Today the company's 600 employees, mostly women, write programs and consult for an international list of clients. About half the workers are linked directly to the home office's central computer.

Implementing a Telecommuting Program

Any telecommuting system has three elements: a centrally located computer that distributes the work to be done and receives the finished product; a terminal used by the employee to accomplish the work; and a telecommunication link—usually telephone company lines—connecting the terminals with the central computer. Technically such a system is little different from those already in use in any large business that has a number of work stations throughout a building connected to a central computer for word processing, data retrieval, or other functions. The decision to incorporate telecommuters into a company's workforce involves much more than hardware decisions. Consideration must be given to three areas: the nature of the work involved, the most appropriate type of work center, and supervision and compensation.

The Kind of Work. Certainly, not every job can be handled by telecommuting. Of the 50 million or so jobs that consist mainly of information manipulation, no more than a third are suitable for conversion outside an office setting—about 15 million jobs in the United States. The most frequently mentioned are computer programming, various kinds of data analysis and input, writing and document creation, and word processing.

The first workers to telecommute were computer programmers. Their jobs entailed sitting in front of a terminal all day and interacting with a computer, something that could just as easily be done miles away from the computer. After programmers, clerical workers usually come to mind. As we enter the era of the electronic office, much of the typing, transcription, dictation, filing, and routine paperwork most businesses generate is being handled by computer. With a terminal to access that computer, or even personal computer disks sent through the mail, clerical employees can do much of the work at home.

Certain kinds of higher-level positions also involve tasks well suited to working at home. Some managers, even those in traditional manufacturing industries, spend considerable time analyzing computer data and writing reports. That time could as easily be spent at home with a terminal. In many businesses the main reason that sales representatives come in to the office is to file reports and do paperwork, which could just as easily be done at home. Some companies are encouraging executives to ask themselves if they really need to come in to the office every day—whether they might spend their time more productively at home, electronically linked to the office.

Remote Work Stations. Mention telecommuting and the first thing that comes to mind is a keyboard and a screen, perhaps also a printer, located in a study, corner of a bedroom, or on the kitchen table. At present this kind of in-home setup represents the most common version of telecommuting. However, two other possibilities exist, also. The first is the neighborhood work station.

Rather than require certain clerical personnel to commute downtown each day, a company can rent office space in a suburban area, perhaps even in a shopping mall. Employees living nearby can drive or perhaps walk this shorter distance to their work stations, maybe even go home for lunch or other breaks. A relatively small office can accommodate several terminals, all linked to the central office downtown.

A second alternative to working at home is the portable work station. Laptop microcomputers can be connected to a central computer from virtually any place that has a telephone. The worker can move freely from one place to another, logging on to the central computer as needed to transmit finished work or receive needed data or further assignments. Workers who need only limited graphic capability now have available portable terminals as small as a paperback book. Portable work stations have great potential for increasing the productivity of any worker who must make special trips in from the field one or two days a week to the home office to do paperwork or search files for information.

Supervision and Compensation. The prospect of having a number of a company's employees working at home—or at least not on the premises—raises some interesting questions for supervisors. The first might be, "How do I know they're actually doing the work I'm paying them to do?" In some cases this problem can be solved by simply keeping track of when and how long a terminal is used. In others the answer is not so simple.

One alternative is to pay for actual tasks completed rather than for hours spent at the terminal: a piece-work arrangement rather than wages. For instance, a programmer might be paid for programs written, not hours spent at the terminal writing; a typist for pages completed. This raises the specter of computer sweatshops and questions about how to handle other parts of a telecommuter's compensation equitably, like health insurance and vacation time.

Potential Benefits

On a day-to-day basis, telecommuting can be advantageous both to employers of large groups of information workers and to the workers themselves. From the workers' perspective, telecommuting means no longer having to deal with the daily drive into town. Using terminals in their homes, clerical workers do not have to spend money and time to dress for work. They have more flexible hours, using their former commuting time to earn more income or for leisure activities. Work schedules can be condensed or spread out according to their needs and preferences. Part-time work is more readily available and easier to take on. Depending on the type of work and the age of their children, many telecommuters can keep the children at home, eliminating or reducing the cost of daycare.

From the employer's viewpoint, telecommuting offers the prospect of reducing the expenses incurred in the operation of large, centrally located office buildings. Businesses that utilize telecommuting find they can operate with greatly reduced overhead and space requirements. For example, with typing pools moved out, floor space can be used for other purposes. Desks, chairs, and other basic work station equipment represent a significant investment for many businesses, much of which are not needed by a telecommuting workforce.

Telecommuting allows employers more flexibility in workload management; word processing and other information-handling tasks need not stop each day at five PM if home workers are willing to take other shifts. Finally, by moving information instead of people, businesses can tap into pools of labor talent not otherwise available. Persons not able to commute, such as the handicapped or those who must stay at home to care for children or other dependents, can be employed on a part- or full-time basis using telecommunication links between home and office. Persons who would not commute to a job that paid on a per-item-completed or piece-work basis might find it worthwhile to take on such work in their homes, giving employers another source of part-time workers. Such advantages have led a number of large businesses to evaluate the possible benefits of telecommuting. But there are potential drawbacks, too, for both employers and workers.

Negative Aspects of Telecommuting

Management Resistance. Despite telecommuting's potential advantages, a person proposing a telecommuting program may encounter resistance from management, particularly at middle levels. For example, a manager directly responsible for the work output of word processing or data input personnel may see any kind of nontraditional work arrangement as undermining her control of that output. Supervisors often feel they lack proper experience and training to deal with employees working at home. Many simply do not feel that they have any immediate problems that telecommuting can solve. Personnel departments may see a telecommuting proposal as needlessly complicating their payroll and record-keeping functions because it creates a new class of employee and generates additional paperwork. For these reasons, most firms venture into telecommuting on an extremely limited basis initially, if at all. Small pilot projects using temporary personnel and heavy restrictions on high-level telecommuters tend to be the norm, rather than dramatic changes. It can take considerable time to overcome the notion that where you work is as important as what you get done.

Labor Concerns. For workers telecommuting does not always represent an improvement over traditional work arrangements. At the individual level, a move from the office to the home can present new problems to replace old. For some volunteers in telecommuting experiments, working at home is a lonely, isolating experience. They miss the camaraderie and social stimulation of the office. Some people gain weight. Many who move home in order to save money on child care find it difficult to get their work done and put their children back into daycare, losing much of the economic advantage they had hoped for. It can also be difficult to convince friends and neighbors that a telecommuter cannot socialize during working hours.

Labor unions can see telecommuting as merely a high-tech version of the sort of work-at-home plans used to exploit workers rather than provide them new opportunities. Many kinds of blue-collar work-at-home programs were outlawed by the Fair Labor Standards Act of 1938 due to management abuse of minimum wage, overtime, and child-labor laws.

At least eighteen states have laws placing restrictions on various types of home work.

Some unions have warned against the development of electronic sweatshops in which telecommuters, constantly under electronic surveillance through their terminals, work without reasonable pay and benefits in order to keep their jobs. They fear that millions of home workers, separated from the day-to-day companionship of their co-workers, will be easy victims of unscrupulous managers. They suspect that telecommuting programs are designed to undermine union solidarity and control. The term *telescab* has been used to disparage part-time telecommuters brought in to replace full-time union office workers. Some unions have placed outright bans on telecommuting. In 1982, for example, the 780,000-member Service Employees International Union banned telecommuting for all members in federal jobs. The Communication Workers of America and the AFL-CIO have also publicly warned of the potential threat of telecommuting to the working standards of clerical workers.

Telecommuting in the 1990s and Beyond

Although growing, telecommuting remains at the experimental level nationally. As the trends mentioned at the beginning of this chapter continue to force employers to seek new employment schemes, this may change. According to various estimates, between 5 and 15 percent of the workforce may be telecommuting by 1990.[5]

Clearly telecommuting is not suitable for every worker or every business. But should the short-term obstacles to telecommuting be surmounted, including resistance from management and organized labor, the coming decades could bring a time when a significant part of the workforce holds jobs performed at or near home, as the nation's economy continues its transition from an industrial to an information base. Looking toward that distant time, Alvin Toffler suggests that telecommuting might have a profound impact on society. If workers own their own terminals in the information society of the future, they would control much of the means of production of information. Consequently, many of today's labor problems and economic issues might disappear, only to be replaced by entirely new ones. Ultimately, the concept of work itself might change.[6]

Notes

1. Conducted by Jack Nilles and a team at the University of Southern California; for a summary and discussion see Alvin Toffler, *The Third Wave* (New York: William Morrow, 1980), p. 216.

2. "Working Mothers Now the Norm, Not the Exception," *Telecommuting Review,* (1 July 1988): 17.
3. Erik Larson, "Working at Home: Is It Freedom or a Life of Flabby Loneliness?," *Wall Street Journal,* 13 February 1985, p. 33.
4. "Penny Plans Another Expansion of Telecommuting Operation," *Telecommuting Review* (1 July 1988): 1.
5. David Scott, "Will a Computer Free You from the Freeway?," *Christian Science Monitor* (29 October 1984): 17.
6. Toffler, *Third Wave,* p. 220.

PART VII

Constant Contact

18

Cellular Mobile Telephones

The cellular mobile phone, like most of the telecommunication systems in this book, is not an entirely unprecedented new communication medium. It is a technical enhancement of mobile radio communication, which has been around nearly as long as radio itself. This chapter reviews how today's car telephone systems evolved, outlines cellular system operation and regulation, and provides a glimpse of cellular development outside the United States.

The "Upwardly Mobile" Radio

Radio communication technology, in one form or another, achieved various modes of mobility not long after radio was first invented. For example, the Wireless Ship Act of 1910 required U.S. freight and passenger liners to have a "wireless" on board for emergency telegraphic communication. Mobile voice communication at sea soon followed, as did portable radios, used in World War I and dubbed "walkie-talkies."[1] After the war the first automotive versions of mobile radio appeared, among the earliest a mobile radio system used by the Detroit police department in the early 1920s.

During the 1920s and 1930s the use of mobile radio spread rapidly to police and other public safety and law enforcement agencies. Private use by taxi, trucking companies, and other businesses with mobile operations quickly developed into the patterns familiar today. All these types of mobile radio are licensed by the FCC and generally referred to as land mobile radio to distinguish them from marine and aeronautical services.

Early Mobile Telephone Service (MTS)

By 1946 mobile radio technology had evolved to the point that a new version was introduced: mobile telephone service (MTS). MTS offered two distinct advantages over regular mobile radio. First, MTS allowed persons using it to send and receive at the same time, a process known as full duplex operation. It did not require the user to hold down a button while transmitting, then release it to receive, as is still the case with conventional mobile radio. Most people are familiar with citizens' band radio, which operates in this manner, referred to as half duplex.

The second advantage of MTS technology was that it made possible permanent interconnection of mobile radios and the local telephone system. Conventional mobile radio communication was, and still is, limited to two persons with radios in their cars or one at a base station. Linking a phone call to conventional mobile radio requires a special procedure called patching. For the first time, MTS allowed routine conversation not only between two persons in cars but also between a person in a car and another using a conventional telephone. The effect was that MTS brought the convenience of mobile radio technology to virtually anyone with a telephone.

The first commercial MTS service in the United States began in St. Louis, Missouri, in 1946. The next year a system was established providing mobile telephone service along a major highway from Boston to New York. Initially MTS was offered only by local telephone companies, with whose telephones the car phones were linked. But in 1949 the FCC set aside additional MTS channels for use by nontelephone companies. This established a dual system of MTS providers in each community: the local phone company or telco and a private competitor, called a radio common carrier (RCC). Each telco was required by the FCC to allow its competitor, the local RCC, to connect its radio service to the local telephone system.

Limitations of MTS and IMTS

Through the 1950s demand for MTS steadily grew, despite the inconveniences of using the system. These included manual channel selection (like CB radio today), difficulty in finding an open line during times of heavy usage, the necessity of placing calls through a special mobile operator, and long waiting periods to subscribe, sometimes a year or more. During the 1960s technical advances eliminated manual channel selection and the mobile operator; calls could now be placed automatically. Signal interference and fading problems were reduced. The new, automated MTS officially was redesignated IMTS, Improved Mobile Telephone Systems, in 1969. However, the most serious problems still remained: the inability to find an open channel when you needed it most and the long waiting lists to subscribe.

Both these problems arose because the FCC allocated to MTS only a small number of channels—forty-four. Ideally, this would permit forty-four subscribers in each city to use the system at any given time. However, interference problems between MTS systems in the same or adjacent cities often reduced the number of usable channels drastically. In New York City, for example, only twelve channels could be used on a major MTS system during peak hours because of potential interference problems. Entering the 1970s subscriber growth in the MTS industry began to level off, simply because there were not enough channels available to permit more growth without further deterioration in the quality of service.

The Development of Cellular Mobile Service

During the late 1960s the lion's share of the IMTS business in the United States was under control of local phone companies owned by AT&T. Seeking to tap the potential market untapped by IMTS technology, AT&T brought its considerable resources to bear on developing a more efficient type of mobile phone service that would accommodate more subscribers. In 1971, in response to the FCC's general request for such proposals, AT&T submitted a plan for a new type of mobile phone service that came to be known as cellular mobile. As explained below, cellular mobile technology greatly increases the number of subscribers a mobile phone system can serve.

In 1974 the FCC set aside for the development of this new cellular service the approximate equivalent of six television channels of spectrum taken from the high end of the UHF television band (channels 70 to 83). Using cellular technology this allocation provided 666 cellular mobile channels per market. After first restricting cellular mobile operations to only local telephone companies, the FCC reconsidered and followed essentially the same telco/RCC dual arrangement already existing in

IMTS. Each city was to have two cellular mobile licensees: the local phone company and a radio common carrier.

Within three years two experimental cellular mobile systems were operating: one in Chicago owned by an AT&T subsidiary called Advanced Mobile Phone Service (AMPS), and one in the Washington-Baltimore area established by American Radio-Telephone Service (ARTS), a radio common carrier. The two test systems demonstrated that cellular mobile was technically manageable and economically promising.

Cellular Takes Off

The FCC formally established cellular mobile radio as a new category of mobile radio service and began accepting license applications in 1982. By then the enthusiasm demonstrated by the subscribers in the AMPS and ARTS pilot systems had generated widespread investment interest in cellular on the part of a variety of communication companies. The FCC was inundated by applications in what *Fortune* magazine called the "cellular gold rush."[2] Using a system similar to that devised to process thousands of LPTV applications (see Chapter 9), the FCC accepted applications in stages, beginning with the top thirty markets. It also set up a lottery system to deal with mutually exclusive applications outside the largest markets.

By 1988 all but three of the top 120 largest markets in the United States had a cellular service licensed. Local phone companies had received licenses in 114 of the top 120 markets, RCCs in 89.[3] Cellular mobile became one of the decade's fastest-growing new electronic media, with two dozen manufacturers offering competing lines of equipment through a variety of retail outlets in most cities. Since 1983 retail prices for cellular telephone equipment have dropped from around $3,000 to less than $1,000, and charges for placing calls have declined somewhat. Cellular has expanded beyond its primary market, the business user, to include government employees and even some private consumers. Industry analysts estimated more than 1.5 million cellular telephones were in use in late 1988 and projected the number would nearly double by 1991.[4]

Cellular versus Traditional MTS Technology

Conventional Car Phones. Perhaps the best way to appreciate the workings of a modern cellular radio system is to begin with a brief description of what cellular replaced, conventional mobile telephone service. A typical IMTS system consists of a single transceiver (combined transmitter/ receiver) mounted at the highest possible point in the geographic center of the service area. The transmitter must be powerful enough to cover the entire service area efficiently by minimizing interference and dead spots where signal strength suddenly drops off, causing annoying disruption of conversations. Subscriber equipment consists of a telephone-like handset connected to a mobile transceiver, usually installed in the trunk of the car, and a rooftop antenna.

The central transmitter is connected by telephone line to the nearest telephone company switching office. Each IMTS system has forty-four channels allocated. Ideally, this allows forty-four subscribers, and no more, to use the system at any given time, each call using one channel to link it with the local telephone system.[5]

The Cellular Principle. In the late 1940s, when commercial MTS services first were being offered, scientists at AT&T's Bell Labs already had envisioned a solution to MTS's limited user capacity and often spotty signal coverage. Bell reasoned that a much more efficient use of the channels in a service area could be achieved by eliminating the single, relatively powerful, central MTS transceiver. Instead, the service area could be subdivided into several mini-areas or cells, each with its own central transceiver. These cellular transceivers and the mobile transceivers in the users' cars could all operate at much lower power because they were closer together. The use of multiple low-power cells would allow subscribers in cells in separate parts of the city to use the same frequencies simultaneously without eavesdropping or interfering with each other's conversations. All the cells would be controlled from a central switching office that was linked to the local phone company.

However, this elegant frequency-sharing solution remained on the drawing board for two decades because of the expense of multiple transceivers and the switching problems they posed. When any user's car passed out of a cell, there would have to be a way to determine where the car was headed, locate an open channel in the next cell, and transfer the conversation to it without noticeable interruption. Furthermore, this tracking and switching operation would have to be done for up to forty-four cars simultaneously, a feat that, at that time, exceeded the practical limits of technology. It took the development of computerized switching and integrated circuitry in the 1960s to solve the switching problem and make multiple transceivers economically feasible.

Cellular Systems: Components and Costs

Components. Today's cellular mobile telephone systems have three major parts: the mobile telephone (in a car or portable), the cell sites, and the central switch. The car phone system includes a handset and cradle of some sort, a keypad, and displays to indicate the system's status. The mobile transceiver is usually secured in the trunk of the car and the antenna installed on the roof or trunk exterior. Portable systems have the same three components, but in a combined, miniaturized form that permits them to be carried by hand or in a briefcase. Generally portable units have less power, and consequently less range, than in-car units.

A typical cell site has two parts: a cell site controller and one or more antennas. Most sites have a single antenna, but multiple antennas may be installed to deal with special signal coverage or reception problems. Antennas usually are installed on towers or tops of existing buildings 100 to 150 feet above ground.

Each antenna is linked to a site controller, which is housed in a small building or shed and consists of computerized switching and monitoring equipment and each site's transmitter and receivers. A site controller relays not only conversations between the cars in its cell and the central switch but also a constant stream of data on separate control channels. This includes information as to which of the cell's channels are in use, user identification, time on the air, and the changing signal strength of each user's transceiver as it passes through the cell. Each site controller is linked to the central switch by private telephone lines, fiber optic lines, or microwave.

The central switch is the brains of a cellular system and the link to the local telephone company. Basically, it is a computer that routes calls between the local phone system and the cell sites, or between two cell sites when a conversation is taking place between two cars. The switch also decides when to transfer a user from one cell site to another, based on the signal strength information coming in from the cell the car is in and the cell it is approaching. This transfer is called a handoff and takes two-tenths of a second or less, making it unnoticeable to the user. Finally, using the data from the site controllers, the central computer switch keeps a record of each subscriber's use of the system and generates billing information.

Costs. In 1988 construction costs per cell ranged from about $500,000 to $1 million. This wide range is due to factors like the price of real estate in urban areas, the cost of tower construction and antenna installation, and the speed and sophistication of the site controller. The number of cells in a system varies with the size of the service area and the amount of traffic: less than a dozen cells are adequate in many smaller communities, while several times that number are required in a major market service area. Central switches typically cost around $2 million. In addition to construction costs, the application process is expensive because the documentation required by the FCC requires the use of technical, financial, and legal consultants. Preparing and submitting an application can cost tens of thousands of dollars, with costs over $100,000 not uncommon for major systems.

When an existing cellular system is sold, the asking price is based on both the cost of the installed equipment and an estimate of future revenues from subscribers. With the average subscriber paying about $125 per month in access fees and time charges, plus another $21 a month in equipment payments in 1988, system buyers were paying up to $100 per subscriber for existing cellular companies.[6]

System Expansion

Once a cellular mobile radio system is established, subscription growth or changing usage patterns may require modification of the system. Most often the requirement is for expanded service in certain parts of the system. For example, the system may simply grow out at the edges, like a honeycomb. Often such expansion is internal rather than external—that is, more channels must be added to certain existing cells where users are having difficulty getting an open line. These channels may be borrowed from underutilized cells.

If new channels cannot be added or existing ones reallocated, then existing cells can be subdivided into smaller cells. This process, called cell splitting, uses a cell's existing channels more efficiently. Each smaller cell has its own transceiver, which operates at a lower power and covers a smaller area than the original cell, allowing existing channels to be used by more cars simultaneously. Thus, a system that starts out with uniform cells in it may evolve into one having different size cells, the smaller cells concentrated in high-traffic parts of the service area.

Cellular Industry Structure Issues

As described earlier, traditional mobile telephone service (MTS and IMTS) has been an industry dominated by the telephone companies as providers. When the FCC established the cellular industry in 1982, it sought to provide competition to the phone companies with its licensing scheme. It set aside one cellular license for the existing local telco in each market and set aside another for which other companies could compete.

The divestiture of AT&T and the deregulation of the nation's telephone industry, however, took place at about the same time that the cellular industry was getting off the ground in the 1980s (see Chapter 13). The new regional Bell holding companies (RBHCs) were allowed to set up various subsidiaries to get into new telephone-related businesses. All of them established cellular radio subsidiaries.

Because these subsidiaries are not telephone companies (even though they are owned by telcos), they have been allowed to compete for the "other" cellular licenses in cities where their parent companies do not offer the regular telephone service. They have also been permitted to purchase other companies already operating nontelco systems in many cities. The result has been that, as of mid-1988, seven of the ten largest nontelco cellular radio firms were RBHC subsidiaries.[7]

This has caused some concern that nontelephone companies seeking licenses will find themselves pushed out of the picture by phone company subsidiaries bidding against them or eventually purchased by them. The result may be an industry controlled mostly by phone companies, either directly licensed in their service areas or indirectly through subsidiaries in cities outside their service areas. Critics point out that this sort of industry structure was what the FCC was trying to avoid in 1982 by establishing two categories of operators in the first place. Others reply that the public is still receiving the benefit of competition in cellular; it is just "Bell versus Bell" competition, not "Bell versus Other."

Cellular Systems Outside the United States

Cellular mobile became a worldwide industry in the 1980s, with services available or under development in most industrialized countries and in the business centers of some lesser developed nations. In Canada, Western Europe, Asia, and the Western Pacific the growth and popularity of cellular telephones has paralleled the U.S. experience. However, major technological and regulatory differences exist compared to the U.S. model.

In countries where telecommunications remains a government monopoly service, cellular is available only through the government phone system. In others, like the United Kingdom and Japan, where privatization of the telephone system is being introduced, cellular mobile services are offered on a competitive basis, private companies alongside the dominant telephone monopoly.

However, to the average business traveler, the most important difference outside the United States is cellular's version of the same technical standards problem that is a theme throughout this book. In North

America, Canada and the United States both use AMPS, but elsewhere there are a half dozen major types of cellular systems in use, all incompatible. In part this is due to different frequency bands being allocated for cellular services in different countries. For example, the United Kingdom uses a derivative of the U.S. AMPS system, called TACS (Total Access Communication System), which is incompatible with U.S. AMPS because it must use different transmission frequencies.

More important, major European and Asian manufacturers and governments developed their own systems rather than lose the market to AT&T or other competitors. This has led to a fragmented world market and lack of service continuity. In Western Europe, France, Belgium, and Germany share common borders, but their systems are incompatible. This makes it impossible, for example, to use a French cellular phone after you cross the Belgian or German border only a few hours' drive from Paris. The European telecommunication ministries are attempting to address this situation by developing a standard for a common system that will be put in place in the 1990s. Elsewhere, Australia, Hong Kong, and South Korea adopted the AMPS system; Japan has its own; and China is developing a system derived from the Scandinavian standard.[8]

Cellular Telephone Service in the 1990s

Most industry observers agree that the upcoming decade holds great promise for this new medium. Continued subscription growth is expected as the cost of equipment and service decline and technological improvements continue. In 1988 pocket-size cellular phones were already available, leading some observers to speculate that in the next century, cellular telephone systems may even become serious competitors for conventional telephones.[9] A major near-term development will be expanded roaming capability. It is already possible to drive from Boston to Washington without a break in cellular service, being handed off from one company to another through special switching and billing systems. Soon, coast-to-coast uninterrupted service along certain major interstate routes may be possible. Furthermore, the cellular systems of the 1990s also will be used more commonly to transmit data from portable computers in cars to homes or offices. Early versions of cellular modems, devices that provide an interface between a portable computer and a car phone, are already on the market.

As cellular phones, in-car and portable, carry a larger portion of commercial voice and data traffic, more sophisticated encryption devices will become available to enhance the security of cellular transmissions. Today unscrambled cellular conversations can be readily intercepted by corporate spies and other eavesdroppers using widely available, relatively simple radio monitoring equipment. One security consultant in the Washington, D.C., area claimed that fifteen of sixteen detective agencies randomly selected from the Yellow Pages offered cellular eavesdropping services.[10]

Finally, in the United States and the rest of the cellular world, manufacturers are already looking toward the eventual conversion of today's analog transmission systems to more efficient digital systems. The all-Europe compatible system mentioned above will be the world's first major digital system. As increased use causes crowding in major cellular service areas elsewhere, the gradual changeover to digital technology will permit new systems to handle three times as many subscribers as today's systems, using the same channels.

Notes

1. *Mobile* generally refers to radios that are attached to a vehicle; *portable* to radios designed to be carried.
2. "Gold Rush at the FCC," *Fortune* (12 June 1982): 102.
3. "An Overview of Cellular Telephone Service," Report #TC27-003, (Delran, N.J.: Datapro Research, April, 1988), p. 101.
4. Herschel Shosteck, "Why Cellular Is Taking Off," *Telecommunications Engineering and Management* (1 August 1988): 29–31.
5. Because car-to-car transmission requires only one channel for two subscribers, technically as many as eighty-eight subscribers could use the system in that way. In practice, interference and dead spots often reduce the number of usable channels to considerably less.
6. For examples of such sales, see Tom Kerver, "Getting Mobile," *Cable Television Business* (1 May 1988): 42–51.
7. See Maribeth Harper, "Will the RHCs Devour the Cellular Industry?," *Telephony* (11 July 1988): 25.
8. Technical comparisons of major world systems can be found in "Cellular Communications," Report #IT-35-120-101 (Delran, N.J.: Datapro Research, February 1987).
9. See William Ginsberg, "The Full Promise of Cellular," *Telocator* (August 1988): 24–27.
10. E.T. Smith, "How to Beat the Snoopers," *Telephone Engineering and Management* (1 August 1988): 100–1.

19
Radio Paging

The familiar pocket pager, which most of us have seen clipped to a physician's belt or heard beeping at a meeting, provides an indispensable communication link for many persons when they are beyond the reach of a telephone. However, like the car telephone, the humble pocket pager today is evolving dramatically. From a simple beeper it is becoming a portable telecommunications terminal, capable of transmitting numbers and text, voice messages, stock quotes, even sports scores to subscribers across the nation. This chapter explains how radio paging services work and the new capabilities and uses we can expect from them in the next decade.

The Birth of Radio Paging—in Hospitals

In 1949 the FCC designated four radio channels for one-way paging services. The first modern radio paging systems were developed for hospitals. The constant clamor of calls over hospital public address systems inspired inventors in the United States and England to devise a silent method of calling doctors to the telephone. Rapidly, such on-site paging systems were improved; their range expanded to include whole cities, and, as the FCC allocated higher frequencies, their ability to receive signals in urban areas improved.

By the late 1950s paging service had become a small but growing industry, with about 100 paging systems owned by radio common carriers

(RCCs) competing with systems owned by telephone companies. The service grew slowly over the next decade amid industry complaints that the channels allocated for paging were not technically adequate. When the FCC made available four higher-frequency paging channels in 1968, the industry began to take off. The demand for pagers was growing, even though the equipment itself was bulky by today's standards; a typical receiver was about the size of a hefty paperback book—not quite compact enough to be rightly called a pocket pager. However, as the new microelectronic technology was applied to paging systems, the receivers became more compact and versatile, and subscribers increased to nearly half a million by the mid-1970s.[1]

The Industry Today

In the early 1980s the paging business hit the channel-allocation jackpot. The FCC set aside twenty-eight more channels usable for traditional paging services. It also made available sixty new high-frequency channels for a new generation of more sophisticated paging services, including national systems.

At this time about 750 RCCs served 80 percent of the paging market; phone companies had the balance. However, after the deregulation of the telephone industry and the breakup of AT&T in the middle of the decade, telcos began to expand their share of the industry by purchasing paging companies. By 1988 telcos had doubled their share of the industry to around 40 percent, with nontelco operators holding 60 percent. More than 6.5 million pagers were served by FCC-licensed telcos or RCCs, with the number increasing rapidly.[2] Thousands of private systems in hospitals, fire departments, law firms, government offices, real estate firms, and so forth added countless other users.

Paging System Components

Receivers. The pager, or receiver, is the familiar device carried by the subscriber. Pagers are simple radio receivers set to only the frequencies used by the paging system. They are addressable, in the sense that a receiver responds to a system transmission only when the signal contains the receiver's unique code. In effect, each pager continuously monitors the paging system's transmission channels, waiting to hear its number. If it hears its number, it takes action. Receivers fall into three main categories, depending on their response capabilities: tone-only, tone-voice, and display pagers.

Tone-only pagers, otherwise known as "beepers" ("bleepers" in the United Kingdom), make up the vast majority of the 6.5 million paging devices in use. Upon receiving their code, they merely emit a series of high-pitched tones. Tone-only pagers about the size and shape of a pen are commonly available. Models that vibrate or flash a light can be used when beeping is not suitable. Once alerted, subscribers must call the paging system operator to find out who is trying to reach them.

As the FCC has expanded the number and capacity of paging channels, more versatile pagers have joined the tone-only pagers in increasing numbers. Tone-voice pagers can, in addition to an alerting tone, transmit a short live or recorded voice message from the operator or even the caller—anything from a phone number to birthday greetings. This allows the subscriber to respond to the caller directly by phone, eliminating the need to call the paging system and find out the number.

Display pagers feature a small screen like a pocket calculator's, which can display numbers, letters, or both. The numbers-only pagers store and display not only the caller's phone number but may include a caller ID number, a call-back priority code, or the time of the call. A single digit can represent up to ten different action codes for the subscriber, if the numbers and instructions are agreed on in advance by the subscriber and caller. Alphanumeric pagers display both numbers and letters, permitting the text of messages to be transmitted and stored. Message length varies from a few characters to hundreds, depending on the storage capacity of the pager. Some versions have miniature printers included that make permanent paper copies of text transmitted.

Transmitters. Radio transmitters broadcast the signal activating the pagers. A paging system may have one or dozens of transmitters, depending on the size of the area served. The number of transmitters needed depends primarily on the size of the area served. Other factors are the frequencies they use, their power level, and physical characteristics of the service area, such as tall buildings or broken terrain. Paging transmission frequencies are in approximately the same part of the spectrum as VHF and UHF television and FM radio. Some systems may also transmit on the unused portion of an FM radio station's signal, called the subcarrier, by arrangement with the station's licensee.

Terminal. The transmitters are controlled by the paging system terminal, the heart of the system. The terminal consists of the electronic devices used to activate the transmitters and provide the interface between the page requests coming in over the telephone system and the transmitters. A terminal might be a box small enough to sit next to a telephone or a number of computerized components requiring a dedicated space. The terminal's capabilities determine the number of pagers and level of service the system can support. Manufacturers classify and price their terminals according to characteristics such as pager capacity, the number of incoming phone lines they serve, the number of transmitters they can control, and the level of automation their computer software offers.

System Operation

The Paging Process. In an elementary system the paging process involves three steps. First, a person wishing to contact a subscriber calls the paging service using a local number over the public telephone system. The paging system operator answers and requests the name of the person to be paged. The system then broadcasts throughout the paging system service area a numerical code assigned only to that subscriber's pager. Although all the pagers in the area will receive the broadcast, only the pager or pagers programmed to recognize that particular code will be activated.

As terminals and receivers have become more sophisticated in the last decade, system operation has become more automated and offers more options to subscribers. Many systems no longer require live operators, for example. Computerized terminal equipment includes software that generates a voice to answer the paging call and directs the caller to enter an ID number for the subscriber using a touch-tone phone. Terminal computers can also page multiple subscribers at the same time, called group page service, or the same subscriber multiple times at set intervals, a service known as repeat page, for which an additional fee is charged. Lastly, automated terminals can also respond to calls forwarded from voice messaging systems (see Chapter 20), notifying the subscribers to call their electronic mail boxes.

Technical and Economic Tradeoffs. The design and operation of any radio paging system involves a number of choices among factors such as system expense versus capabilities, and capabilities versus channel capacity. Generally the more information a pager conveys, the more expensive it is and the fewer subscribers the system can handle at any given time on a channel. For example, in some systems as many as 30,000 subscribers are served using a single channel and tone-only pagers, but the only information being transmitted is the beeper code signal and only for a fraction of a second. Display pagers, which transmit many times for information, reduce the number of possible users substantially. For example, an early model of the Motorola Optrx pager could display messages up to eighty characters long but could serve only half as many subscribers per channel. Because of the amount of bandwidth required for voice transmission, tone-voice pagers are the least efficient: only about 1,200 can use a single paging channel.

Tone-only pagers sell for about $100, though that price is dropping. Pagers can also be rented annually for about half the purchase price. Typical subscription fees run $10 to $20 a month. Display pagers, on the other hand, cost considerably more. The Optrx mentioned above sold for $500. Thus a commercial paging system operator must determine what mix of receiver types and subscription fees will maximize subscribership, while balancing that against the greater investment required by sophisticated terminals and receivers.

Expanding the Reach of Paging Systems

Industry observers often categorize paging operations according to their reach or coverage. The smallest in area are on-site systems, designed to cover a single building or building complex. Examples include private systems in hospitals, factories, or office complexes. On-site systems typically have only one transmitter. When paged, the employee uses a house phone or intercom system to check in with the system operator, who is often the operator of the organization's telephone switchboard.

At the next level are wide-area paging systems, covering entire communities or metro areas, using dozens of transmitters to serve thousands of subscribers. Today paging systems in separate cities increasingly are being linked to form regional paging networks, allowing subscribers to be paged locally even though they are out of town. A major technical stumbling block for this kind of interconnection is lack of equipment compatibility among systems operated by different licensees in different communities. Telocator Network of America, the leading paging and cellular radio trade organization, has led the industry in development of technical systems and standardization efforts to surmount this hurdle.[3]

National paging systems are also being developed. Three companies were licensed to offer national paging services in 1984 after the FCC allocated channels for this type of system two years earlier. One of these applicants, National Satellite Paging, has established a system that can automatically page subscribers in any of eighty U.S. cities.

A person wanting to page someone in another part of the country calls the paging company's national office in Washington, D.C., using a toll-free number. The paging system computer in Washington relays the request over a phone line to a satellite uplink facility in Mountain-view, California. The paging signal is relayed via WESTAR IV satellite to eighty cities that have National Satellite Paging's reception and transmission equipment installed. The designated pager alerts the subscriber with a numeric display of the caller's phone number. The whole process takes only seconds and is handled entirely automatically. A voice messaging service is also provided by this system, allowing the caller to leave a recorded message for the subscriber, who can dial in and hear it after being paged.

The Future

As the message capacity of paging systems increases, so does the variety of information services they can provide. High-capacity automated terminals allow system operators to add services beyond simple paging and personal message transmission. In some metro areas the latest stock market information is available on pocket receivers. Sports scores and game summaries can be monitored on pagers in some areas as well. The humble pager is well on its way toward evolving into a pocket information terminal.

This evolution will be accelerated by digital transmission systems. In paging, digital technology makes it possible to squeeze much more data into a paging channel and deliver it at a much higher speed. By encoding caller messages digitally, transmitting them at higher speeds, and then converting them back to audio signals again, tomorrow's paging systems will make tone-voice and display pagers much more efficient and readily available. Digital systems also will make possible the transmission of entire pages of data to pocket printers, so emergency medical personnel, construction supervisors, and others who need it can have printed data on hand immediately—wherever they might be.

In addition to one-way services, two-way message services are on their way. A nationwide two-way message service planned by Dataspeed uses a pocket device that not only accepts paging signals but has a calculator-size keyboard and can store, send, and receive messages up to 500 words long. Rather than conventional paging channels, the device will transmit and receive on FM subcarrier frequencies.

New services are just the beginning. Any paging signal that can be relayed by satellite nationally can be relayed internationally. In the 1990s paging systems may be able to reach you almost anywhere in North

America or Western Europe. Radio paging is by no means solely a North American communication medium. One of the world's oldest and largest manufacturers of paging equipment is Multitone, a British firm that has been in the business since the early 1950s.

Most of the state-run telephone systems in Western Europe offer paging service today using equipment built by Multitone or its U.S. or European competitors. Public and private systems are also in operation in the industrialized nations of the Mediterranean, Persian Gulf, and Western Pacific. According to Multitone, one of their systems has even been installed in Moscow.[4]

Setting aside nagging compatibility issues, it is clear that global paging is technically possible using satellites. At this point, technical possibilities may exceed the political realities of international communication, and international paging may join DBS and other technologies discussed in this book whose future development eventually will hinge solely on international cooperation.

Notes

1. For a detailed overview of the early development of the industry, see David Scott, "Journey through the Past," *Telocator* (August 1988): 38–43.
2. Jim Bowen, "Opportunity Beeps for Telco Paging," *Telephony* (30 May 1988): 31–35.
3. See Jay A. Moskowitz, "Paging Networks Come of Age," *Telocator* (August 1988): 14–23.
4. Bill Hollins, "The Unlimited World of Paging Systems," in *Developing World Communications* (London: Grosvenor Press International, 1987), pp. 236–37.

20

Electronic Mail: Voice Messaging

Cellular mobile telephones permit conversation on the move, and paging services alert subscribers that someone is trying to reach them. However, neither of these technologies entirely solves the problem of leaving a message for a person who is away from the phone (office or car) or the problem of transmitting a priority message to twenty such persons at the same time. Enter voice messaging, the newest and least well-known technology in the "constant contact" trio. Using voice messaging systems, organizations in the United States, Japan, and Europe are dramatically improving internal communications and productivity. This chapter provides a brief general overview of electronic mail and a somewhat closer look at voice messaging, the newest addition to the array of new media finding their way into everyday business use.

Varieties of Electronic Mail

Voice messaging is only the most recent development in a general trend toward combining computers and telecommunications in ways that allow organizations to communicate more effectively within the same building or across continents. *Electronic mail* generally refers to systems that transmit written communications or graphics over telecommunication lines. Much effort has been focused on developing reliable, cost-effective means of transmitting documents instantly from one location to another, when postal or courier service is not timely enough.

The first electronic mail system was, of course, the telegraph. Today the traditional telegraph message has virtually disappeared from business communications, replaced by three kinds of systems.

Telex. The oldest, Telex, first became available in the 1930s. Similar to the telegraph, Telex offered the advantage of allowing two-way real-time communication between sender and receiver using keyboards. Today Telex is still a widely used means of international communication, with some 2 million subscribers around the world. It combines the immediacy of the telephone with the advantage of a hard-copy record of the transmission.

Fax. Another commonly used electronic mail technologogy is the facsimile system or fax. Fax allows document transmission without the need to enter information by keyboard; pages are simply scanned line by line, transmitted, and reproduced line by line. Fax provides a means of sending not only documents but diagrams or photographs from one place to another over conventional phone lines quickly without resorting to mail or courier service. Since the mid-1980s the decreasing cost and size of fax machines have resulted in the increasingly common sight of the personal fax on office desktops. "Smart" photocopiers and electronic printers are also available today with fax capability.

The proliferation of fax technology has led to an interesting information age problem in many offices: junk fax. Fax machines that are left in the *on/receive* mode throughout the day are kept busy reproducing advertising and other junk mail. To get control of their systems, some organizations have resorted to unlisted fax numbers or coding systems that permit the reception of authorized materials only.

Computer Mail. Using personal computers or terminals linked to larger computers, workers in the same building or in different countries can carry on an electronic conversation on their screens or leave messages for individuals or groups. Computers in the same or adjacent buildings are linked using two somewhat different approaches: local area networks or telephone switchboards.

A public branch exchange (PBX) is a switchboard used to interconnect all the telephones in a large building and link them with outside communication lines. Switchboards in hotels are a familiar example of PBXs. In recent years PBXs have evolved from simple, manual plug-in systems to "intelligent" switching systems that can not only handle voice communications but automatically route digital signals from office computers and other equipment such as fax machines and electronic copiers. These third-generation digital PBXs are one method used to link computers into an electronic mail system.

Somewhat similar to PBXs are local area networks (LANs), which are installed independently from the phone system and usually do not handle normal voice traffic. Because they are specialized data transfer systems with higher capacity than phone lines, LANs can move large amounts of data more rapidly than PBXs. Another advantage is that if the phone system goes down, the LAN may remain in operation. They also require the installation of a new wiring or cable system, however, next to or on top of the existing phone system.

Once the problem of local computer interconnection is solved by a LAN or PBX, the next step in computerized electronic mail is to provide links between offices across town or across the nation. Here the choices fall into two main categories: to use an existing local or national phone system or to bypass it. Sometimes, for economic or other reasons, it is appropriate for organizations to build their own telecommunication systems using terrestrial microwave, satellite, and even fiber optic technology. However, in most cases communication links are supplied by existing service companies. These include the major long-distance and regional phone companies as well as smaller companies called specialized common carriers (SCCs). SCCs construct lines only along certain high-use routes, like New York–Philadelphia or St. Louis–Chicago.

Other services, called value-added carriers (VACs), lease lines from the major carriers and, using their own equipment, add services that may not be available from the phone company. For example, perhaps an organization wants to establish an electronic mail or data link between computers in two branch offices, but the two computer systems speak different languages. A VAC can set up the link routing the data through its own computer, which translates for the other two and allows them to talk to each other without expensive modifications.

In summary, Telex, fax, and computer mail can significantly enhance the efficiency and productivity of most organizations by allowing the instant transfer of documents and graphic information. However, there are times when only voice communication by telephone can provide the immediacy, personal impact, subtlety, or flexibility needed to do the job. Unfortunately, the telephone is often a very inefficient communication device.

The Telephone: Ubiquity versus Usefulness

The global telephone network has been described as the biggest, most complex machine ever built. Without it, contemporary life could not exist as we know it. However, despite its well-deserved status as a technological wonder, as a communication system the telephone network obviously has its limitations—some technical, some human. Although it enables us to place a call spanning the globe, the odds are about three to one against our actually contacting the person we are trying to reach on the initial attempt. Research indicates that fewer than 25 percent of business telephone calls reach their intended party on the first try. Most don't get through due to a busy signal, no answer, or technical problem. If a call does get answered, the party being called is, more often than not, out of the office or otherwise unable to speak.[1]

We then leave a message requesting a return call. The return caller, in turn, confronts similar problems getting through. Another message is left and another round of "telephone tag" is under way. If, as is often the case, either call requires an additional call to a third party for information, the difficulties compound geometrically.

Hard-working receptionists, answering machines, and professional answering services can help, but they do not eliminate missed calls and telephone tag. Consequently, while telephone service links virtually everyone in the developed world, the basic telephone is not yet considered an ideal means of transmitting high-priority information quickly throughout most organizations. Voice messaging, the newest kind of electronic mail technology, provides a remedy for this problem.

Voice Messaging Technology

Essentially, a voice messaging system replaces an ordinary answering machine with a computer, resulting in a much greater variety of features and greater capacity than any answering machine can offer. Messages are stored by the computer and forwarded to other subscribers automatically. Subscribers can call in and check their electronic mailboxes twenty-four hours a day and hear the complete message left for them in the sender's voice.

A typical voice messaging system is contained in a cabinet that may take up as little space as a personal computer or as much as a refrigerator, depending on its capabilities. Regardless of size all systems contain the same essential components: (1) an interface to connect the device to the in-house and external phone systems, (2) a codec (coder/decoder) for transforming voice signals into computer data and back again to voice, (3) a central processing unit (computer) for managing and routing messages, and (4) a storage unit for saving them.

Using a voice messaging system to leave a message works as follows. Using ar ordinary touch-tone phone you dial the system's phone number. Much like an ordinary answering machine, the system's computer answers the call with a recorded reply, listing service options you can select by pressing certain numbers on your phone. You press the "leave a message" number. The computer requests the "box number" of the person or people you wish to contact, and you punch in that information.

At the computer's prompt you speak your message, which can be much longer than the usual answering machine allows. The sound of your voice is digitized by the codec, and this data is stored on magnetic disk. Once you have completed your message and have hung up, the computer sends a "message waiting" signal to the telephone of the person you wanted to reach. This signal may be a flashing light on the telephone set, a special tone audible when the phone is in use, or a ring back, in which the system calls and tells the party a message is waiting.

When the message recipient dials into the system to retrieve the message he or she must enter the correct personal code to have access to the electronic mailbox containing it. The message is recalled by the computer from disk, reconverted by the codec, and played back in its original form.

Typical System Features

In addition to the fundamental task of leaving a message for later playback, most voice messaging systems offer a variety of capabilities designed to reduce telephone tag and enhance productivity. Messages can be sent to one person's box, to several persons' boxes, or to a group of people on a designated mailing list (called a broadcast function), all without having to speak the message more than once.

Messages one person receives can be automatically forwarded from that box to another and to paging services or mobile phones. Automatic answering features allow the user to leave a personalized "I'm not in at the moment" greeting for callers with instructions on how to leave a message or be transferred to a live operator. Messages can be coded to make them confidential and prevent intentional or accidental eavesdropping. On recall, messages can be scanned or previewed, eliminating the need to listen to each in full in the order it came in, as is often the case with conventional answering machines.

Stand-Alone Systems versus Service Bureaus

Any organization considering using voice messaging technology must decide whether to purchase an in-house system of its own or to subscribe to services from a voice messaging company or service bureau. Each option has positive and negative aspects.

Purchasing a system can mean a significant cash outlay, particularly for larger firms. Fully elaborated systems that handle several hundred users cost upwards of half a million dollars. Systems accommodating under 100 users start at around $20,000. Economy systems, targeted at smaller operations, are available for as little as $10,000, including systems that are integrated with standard personal computers.

Installing a large in-house system calls for financial and personnel commitments beyond the initial purchase outlay. Large systems require maintenance and technical monitoring. A training program and user support services often are needed to maximize the number of users. As usage patterns develop, the capacity of the system may need to be increased or reduced. As new features become available, upgrading the system will require additional investment. There may be unanticipated delays or difficulties during the start-up phase of operation or downtime for repairs and upgrading.

On the other hand subscribing to a service bureau permits a quick start-up and immediate access to new service improvements. If service is not satisfactory or voice mail doesn't appear to meet your company's needs, you can switch bureaus or drop voice mail altogether.

Subscription services also have their drawbacks. Although using a service bureau allows a comparatively low-cost start-up, service bureau charges can add up rapidly, typically $10 to $35 per month per electronic mailbox, plus charges are incurred for the use of the bureau's long-distance, and sometimes local, telephone lines. If voice messaging catches on in your company and becomes heavily utilized by a large number of employees, subscription services can be more expensive than buying your own equipment. Other comparative negatives associated with bureau service are less control of the system, including less security; the possibility of overload of the system by other subscribers; and potential difficulty combining voice messaging services with existing communication systems such as PBXs or text messaging systems.[2]

On balance, the basic subscribe-versus-install dilemma is often resolved in terms of economies of scale: the larger an organization and the more use the voice messaging system receives, the more attractive the purchase option becomes. One service bureau suggests to its customers that when they expand their usage to more than fifty voice mail boxes, they should consider purchasing a system.[3] Industry observers also suggest that new users, regardless of size, first try a subscription service. If purchase later becomes a consideration, the decision can be based on first-hand experience with voice mail within the organization.

Notes

1. James Martin, *Telematic Society: A Challenge for Tomorrow* (Englewood Cliffs, N.J.: Prentice-Hall, 1981), p. 90.
2. For a detailed comparison of subscription versus purchase, see Arthur M. Rosenberg, "Voice-Messaging Services or an In-House System: Here's How to Tell Which Is Better for You," *Communication Age* (April 1985): 38–43, or Donald H. Van Doren and Douglas Gold, "Does Voice Mail Address Your Every Wish?," *On Communications* (June 1985): 7–9.
3. David A. Perry, "Voice Mail: The Phone Made More Efficient," *Focus* (18 February 1987): 86–91.

PART VIII

Conclusion

21
Summing Up

Future historians will certainly look on the last two decades as one of the most important periods in the history of electronic media. Virtually every new medium or technology discussed in the previous chapters has undergone a critical development or change during these years, whether technical, regulatory, or economic. In many cases those changes may well have set the stage for even greater transformations in the next two decades.

We have looked at these developments in six areas, which this chapter attempts to summarize: the rise of satellite communications, the transformation of television viewing and broadcasting, the expansion of wireline media, the development of electronic publishing and retailing, the onset of electronic travel, and the increasing reach of personal telecommunications. In closing, the international scope of these developments is reemphasized and the paradoxical nature of technological innovation suggested.

The Rise of Satellite Communications

Domestic communication satellites are at the heart of much of the technological, regulatory, and economic change experienced by the telecommunications industry in the last two decades. In 1963 to 1964 three Syncom satellites were the world's first to be positioned 22,300 miles above the equator in geosynchronous orbit. In their parking spots the Syncom birds permitted trans-Pacific and trans-Atlantic telecommunications using, for the first time, stationary receiving dishes instead of much more elaborate tracking antennas, and were available for use around the clock. They were the prototypes for all the geostationary satellites that followed.

Satellite communications developed on three levels in subsequent years. Beginning with Intelsat in 1964 the first level of development for geosynchronous satellite telecommunications was international. Intelsat today now serves over one hundred member countries with sixteen geostationary satellites around the globe. Intelsat was joined by the smaller USSR-sponsored cooperative, Intersputnik, in 1971, and by INMARSAT, a global maritime communications cooperative, in 1979.

The second level of development, domestic systems, began in 1972 with Canada's Anik, the world's first geostationary communication satellite for domestic service. In 1972 the United States became the first country to permit satellite construction and operation by the private sector under the FCC's "open skies" policy. Today domestic birds make up the single largest category of communication satellites in orbit over the equator worldwide. The third, intermediate, level of satellite ownership, regional systems, has been a more recent development of the 1980s. ARABSAT, EUTELSAT, and PALAPA are examples of regional cooperatives.

The rapid development of satellite communication technology quickly led to demands for parking spaces and transmission frequencies. The global allocation of parking spaces and frequencies is accomplished through the International Telecommunications Union and the government regulatory agencies in ITU member countries. The ITU has established four service categories, of which FSS, fixed satellite service, is the most common type of satellite operation globally and involves the relay of voice and data between a limited number of transmission and receiving sites, such as telephone companies. While satellites have made national and even international communication much cheaper, they do have some distinct limitations. These include transmission delay, antenna size, security, frequency crowding, and meteorological interference. Consequently it is not anticipated that satellite technology will soon replace terrestrial microwave in all instances, and in fact, it can expect competition in some applications from high-capacity fiber optic transmission methods.

At present the demands for parking slots on the Clarke Belt have highlighted its importance as an international resource and the need for equitable solutions to the problems encountered in its use by less developed nations. Ultimately the full potential of satellite communication technology will be actualized only through international political and economic cooperation.

Satellite-to-home video is a specialized use of geostationary satellites, falling under the ITU designation of broadcast satellite service (BSS) or direct broadcast satellite (DBS), depending on the power of the satellite. At present the most advanced development of satellite-to-home video can be found in the United States, where about 2 million backyard dishes have been installed to view program services otherwise available only through cable systems. These dishes pick up the same signals received by cable systems.

The home dish industry in the United States is a forerunner of worldwide DBS. True DBS uses a much more powerful satellite, allowing for much smaller (as little as 15 to 24 inches), less expensive dishes. A French–German consortium launched the world's first DBS satellites. The first, in 1987, was abandoned due to technical failure. The second, launched in late 1988, was scheduled to become operational in 1989. Others are planned for launch then and in the early 1990s in Western Europe, the United States, and Japan. Although the ultimate financial and technical success of these efforts is uncertain, their fruition would be a major milestone in the progress of satellite communications, making satellite reception nearly as convenient as tuning in a local television station.

Transforming Television Broadcasting and Viewing

Until the 1970s the average home television receiver served much the same purpose it did when it was introduced to the public in the late 1930s; to receive broadcast television programming from local stations, most of which served as local distributors for national networks. In recent years that traditional arrangement has been altered by the appearance of a variety of new programming sources, transmission methods, and technologies.

The introduction of a practical home video recorder in the late 1970s changed the familiar TV's use forever. It allowed the viewer to control the time programs were viewed and provided an alternative to conventional programming sources: the video rental store. At the same time the cable industry, rejuvenated by satellite-distributed pay-TV revenues, grew dramatically from 1975 to the present, adding further viewing options for the audience. Today network TV viewing is still the single biggest draw on American TV, but the audience share of the three commercial networks is only about two-thirds of its traditional size and still shrinking.

Recent years have also seen the deregulation and introduction of alternative forms of television transmission. Low-power television is gradually taking its place in the broadcasting hierarchy to meet the need for small-town television and specialized or ethnic programming in urban areas. UHF stations are free to scramble their signals and sell premium programming to subscribers. Microwave distribution of pay programming (MMDS) offers a further alternative to conventional broadcasting and cable in some cities.

The 1990s will see the availability of the next generation of television receivers: high-definition systems with twice the resolution currently available. Japan introduced the first working systems in the late 1970s. Put on the defensive by Japan's plans to market consumer HDTV systems a decade later, the United States and Western Europe are rushing to develop their own working versions in an effort to prevent their current national broadcasting technology from being rendered obsolete by foreign equipment. In the United States the broadcasting networks and their affiliated stations are faced with a prospect unthinkable only a few years ago: becoming the low-tech version of television, the video counterpart of AM radio.

The next decade will also see the impact of laser-optical video systems on conventional TV. The laser disc's enormous storage capacity will bring an entirely new generation of programming alternatives into many homes, providing new interactive uses for the television receiver. Elaborate video games and databases featuring moving images and sound (such as audio/video/text versions of encyclopedias) are already in various stages of development.

The Wireline Media in Transition

Twenty years ago most residences had a single wire acting as an information conduit into and out of the home. It belonged to the telephone company. Today that wire has been joined, on the poles and in half of U.S. homes, by a second wire owned by a local cable company. Both wires, and the industries that own them, have experienced sweeping changes in the last twenty years.

Two decades ago the telephone meant much the same thing to subscribers it had for almost a hundred years: voice communications. The telephone company meant AT&T, which provided virtually all long-distance service and, through ownership of the major regional telephone systems, controlled most local service as well. The 1970s saw legal and political challenges to AT&T's dominance lead ultimately to the breakup of AT&T and the restructuring of the telephone industry. Technological advances have resulted in the provision of enhanced data and video communication services by telephone companies, including teleconferencing.

Two decades ago only about one American household in ten subscribed to cable. Cable service barely existed in major urban centers, and everywhere programming was limited mainly to a dozen or fewer channels carrying local or regional television station signals. The 1970s, however, saw the development of satellite program distribution and consequent rebirth of the cable industry, which included its penetration of densely populated urban markets and surrounding suburbs. Federal legislation in 1984 gave cable operators freedom to set prices according to local demand for services and removed much of the threat of franchise renewal proceedings. Today cable penetration approaches 60 percent nationally and most subscribers have available twenty or more channels of satellite distributed network programming. While the telephone system has evolved from a voice system into a telecommunication system, the cable system has simultaneously evolved from a mere extension of broadcast television into a programming supplier in its own right.

Traditionally, these two industries have seen themselves as being in clearly separate businesses: one-way video distribution and two-way voice and data transmission. However, the changes of the last two decades increasingly have blurred that boundary. As early as the late 1970s the cable industry began experimenting with two-way services on new state-of-the-art systems like Warner's Qube system in Columbus, Ohio. Such systems allowed subscribers to communicate upstream to the system headend. Although early attempts at interactive programming eventually failed, two-way cable still exists in the form of automated pay-per-view technology, security systems, and some interactive videotext and shopping services. Cable also emulates the telephone company through its limited number of active institutional networks, specialized cable systems built by franchise holders to provide video and data communication links for government and business organizations in their communities. In a limited way, the cable industry has demonstrated it has the technical capability and economic resources to offer a number of two-way services.

Likewise, with the newly independent regional Bell operating companies free to pursue many lines of communication services formerly closed to them before divestiture, the telephone industry has made some headway into video distribution. Although still barred from outright cable system operation in the late 1980s, the telephone industry was pressing regulators to allow telephone company-owned fiber optic systems to become carriers of third-party video services. The video gateway concept allows the phone company to provide the wire but not to own or control the programming.[1] The prospective development of an international ISDN system in the next decade would further blur the traditional separation of cable and telephone services by making the phone system of the future capable of virtually any kind of two-way information transfer, including video.

Electronic Publishing and Retailing

New electronic media are providing new ways for retailers to distribute information and consumer goods to the public. Several forms of electronic publishing have appeared in recent years, falling into two broad categories: online and offline. Online or videotext services are continuously updated electronic databases that can be searched remotely by subscribers. The first appeared in the early 1970s and were accessible only by trained terminal operators in libraries and research organizations. But within a decade the proliferation of personal computers brought with it databases like Compuserve and the Source, designed and priced for the general public. Today in the United States, Canada, and much of Western Europe, business and consumer-oriented databases are widely available. However, a personal computer must be used as the access terminal, restricting the market for these services to persons with computers in their homes or offices and the inclination to use them for that purpose.

Attempts to broaden the market for videotext beyond computer users/owners—that is, to make it a true mass medium—have met with mixed results. In the United States major trials in Miami and Los Angeles using keyboards and regular TV sets failed to generate significant consumer interest. England's Prestel system began in 1979 as a consumer-oriented service but has succeeded primarily with business subscribers. In France, however, Minitel has been a major consumer success, with some 3 million subscribers anticipated by 1990.

Projects emulating Minitel are being explored in the United States by several regional phone companies. Teletext—online text and graphics transmitted via conventional television signals—has yet to develop beyond the stage of small-scale market trials in the United States, though a number of services exist in Europe, including Britain's pioneering CEEFAX/ORACLE.

The second major category of electronic publishing, offline services, is much newer than online/videotext services. With offline services, the user buys the electronic database and accesses it using special equipment on his or her premises. The data is not continuously updated, though newer versions or supplements can be purchased at later dates. However, there are also no additional charges for using the database beyond the purchase price.

The development of laser disc technology in the late 1970s made possible storage of enormous amounts of text and graphics on disc. Today laser disc compilations of information that has a reasonably long shelf life, such as encyclopedias, various directories, and indexes, and statistical databases, are increasingly available in libraries for public use, replacing microfilm or paper versions of these information sources. The development of a consumer market for these products awaits the appearance of an inexpensive multipurpose laser disc player with a control system that will accommodate text and graphics in addition to today's video and audio entertainment disc formats.

In addition to creating new forms of publishing, new technologies are being applied to traditional retailing methods, creating video retailing systems. For a century, the mail order catalog or telephone was the only means of making a purchase or other transaction without visiting a retailer in person. Today most consumer videotext systems offer shop-at-home services in addition to database browsing. On-site videotext systems in malls and stores featuring self-service catalog information and ordering are becoming common. An alternative approach is being attempted by the national retailer J.C. Penney through its Telaction service, accessible only through local cable systems.

Telecommunication Substitutes for Travel

In the 1970s several trends converged that led to widespread interest and subsequent investment in teleconferencing and telecommuting technology. Domestic communication satellites reduced the cost of many forms of long-distance communication, telecommunication equipment declined in price and increased in capability, digital and fiber optic technology increased the capacity of existing phone systems, and government deregulation of the telephone industry opened up competition in many types of telecommunication products and services. At the same time the cost of transportation rose in the mid-1970s, becoming a significant expense for the first time for many large companies and government agencies. These trends, added to a general concern about declining levels of productivity and competitiveness in American and European industry, created interest in using electronic media as substitutes for business travel.

From corporate computer conferencing to international videoconferences, teleconferencing has become commonplace in the last two decades. The sale of teleconferencing equipment and provision of electronic meeting services has become a multimillion dollar industry in the United States. Major hotel chains offer teleconferencing facilities and services to attract and hold corporate clients. Other companies specialize in producing national or international meetings for clients. However, in-house two-way videoconferencing facilities are still restricted to the few large organizations that can justify their expense. Although experience and research have confirmed that teleconferencing is not suitable for every type of meeting, many varieties of electronic meetings are becoming common business practice in the United States and other industrialized countries. As concerns about productivity and transportation costs are not likely to diminish soon, long-term prospects for continued expansion of teleconferencing seem excellent.

At the same time a number of organizations are demonstrating that similar technologies also can permit certain types of employees to work at home, an electronic connection to their organizations substituting for the daily commute. Telecommuting programs, though still on a small scale nationally, have not only saved money for employers and employees, but in some instances have opened up job opportunities for persons not able to physically commute or maintain a conventional work schedule because of other obligations.

Telecommuting plans vary considerably. Some telecommuters are full-time employees with regular benefits; others work as part-timers strictly on a wage basis. Some use company-owned equipment, others must buy or rent theirs. Some companies hire only clerical or data-entry workers as telecommuters, others permit managers and other upper-level employees to do so on an occasional or even full-time basis.

Unlike teleconferencing, telecommuting has generated some controversy. In the extremes it is either considered by management to be a threat to supervision and control of its employees, or feared by labor as a management scheme to isolate workers and deprive them of fair wages and benefits. The future development of telecommuting as an alternative to "being there" in many kinds of jobs depends to a great extent on whether these misgivings and suspicions can be overcome.

The Increasing Reach of Personal Telecommunications

It is ironic that new technologies have made it possible to transmit a telephone conversation or television picture virtually around the world but until recently most people in an automobile or taking a walk or in a closed meeting could not be reached directly by any of this impressive array of modern telecommunication systems. Although paging systems and mobile phones were first developed nearly forty years ago, it took the technical advances of more recent years to lower their price and increase their availability and appeal beyond a small cadre of users.

After nearly a decade of development and experimentation, cellular mobile radio was established as a new category of mobile radio service and licensing begun in 1982. Within six years all but three of the 120 largest markets in the United States had a cellular service licensed. At the same time retail prices for cellular telephone equipment declined to about a third of their initial level, though line charges have dropped only slightly. Cellular has become more than a telecommunication system for elite business users and increasingly is seen in the cars of government employees, company sales representatives, construction supervisors, and even a few private consumers. More than a million had been purchased by late 1988, and that figure was expected to double by the early 1990s.

At the same time cellular radio has become a worldwide industry, with systems operating or under development in most industrialized countries and even urban business centers in some less developed nations. However, the United States remains the world's largest unified cellular market because major technological and regulatory differences prevented a single compatible system from developing in Europe. European telecommunication ministries hope to remedy incompatibility and other problems in the 1990s with an all-Europe digital system. In the United States continued subscription growth and technical advances, such as pocket-size phones, national roaming, and much higher-capacity digital systems, have led to speculation that in the next twenty years cellular systems may even begin to compete with the traditional wired telephone system.

When the FCC first made additional paging channels available in 1968, the industry began to take off, even though a typical receiver was about the size of a paperback book—not really a pocket pager by today's standards. As microelectronic technology made receivers more compact and useful, total subscribers increased during the 1970s. In the early 1980s the FCC set aside more paging channels, further accelerating industry growth. By 1988 about 6.5 million subscribers to paging services were divided about 60 to 40 percent between RCCs and local phone companies. Most major state-run telephone systems in Western Europe provide paging service using equipment built by U.S. or European companies. As with cellular phone service, public and private paging systems exist in urban centers and large organizations throughout the world.

Compared to their counterparts twenty years ago, today's paging systems are much more automated, often featuring computer systems that answer calls and page subscribers. No longer limited to individual cities, paging is becoming regional through the interconnection of systems in adjacent markets. National paging systems are also under development.

Paging technology is being used to provide new kinds of data transmission services, including stock market quotes and sports scores. The next generation of paging systems will be digital, permitting many more subscribers to receive a variety of data at higher speeds. The pocket pager at that point will be evolving into a pocket information terminal. Some will be capable of transmitting as well as receiving several pages of text or other information. Ultimately international paging will become technically, if not economically or politically, feasible.

Lastly, computer mail and facsimile terminals allow instant transfer of documents and graphic information. There remain certain situations in which the impact and flexibility of voice communications is desired, although telephone "tag" often makes voice communication by telephone difficult. Voice messaging systems are doing for telephone communication what fax has accomplished for text and graphics.

Voice messaging or voice store and forward systems replace an ordinary answering machine with a computer, allowing messages to be recorded and forwarded automatically from one user's electronic mailbox to another. Users can call in twenty-four hours a day and hear the complete message left for them in the sender's voice. Voice messages can be sent to a single person or a firm's entire sales force in the field with one procedure. Misplaced "while you were out" memos and rounds of telephone tag are eliminated. Combined with cellular phones and automated paging systems, voice messaging systems are closing the final gap in the telecommunications network, allowing persons desiring so to maintain constant contact.

New Electronic Media: An International Phenomenon

Perhaps more so than in past decades, today the innovation, development, and consumption of new electronic media are an international phenomenon. Two important aspects of this are the challenge to U.S. telecommunications preeminence and the worldwide diffusion of certain new media.

Although the United States traditionally has been the preeminent nation in the telecommunications arena, this no longer entirely applies in the case of the new electronic media. Old international patterns of innovation, development, and consumption are giving way to new ones. One important development along these lines in the last twenty years has been the growth of electronic media in Western Europe. For example, there are now more television households in Western Europe than in the United States: about 120 million versus 90 million. In addition, with the ongoing relaxation of traditional trade restrictions between EEC members, deregulation of television broadcasting, and growth of cable and satellite services, Western Europe promises to become a much more important world showcase and consumer of new electronic media market in the 1990s.

Consequently, some new media are taking hold in Europe while their development lags or founders in the United States. For example, the world's first consumer DBS satellite was launched in Europe in 1988, with others scheduled to follow in 1989 and 1990. Videotext and teletext originated in Britain, and after aborted U.S. efforts, the world's most successful consumer videotext system, Minitel, was developed in France. Several U.S. regional phone companies are considering emulating that system. France also has under development the world's largest and most sophisticated fiber optic cable television and interactive consumer information system.

Elsewhere, the world's only version of HDTV ready for market is Japan's NHK system, which has already been accepted by U.S. industry groups as the standard for high-definition studio television production. Japan may also develop a commercial DBS system before the United States.

After years of being solely the province of the United States and USSR, launching of communication satellites is becoming a more competitive international business. The United States and USSR, which now has to offer the world's heaviest rocket, the Proton, and a fledgling space shuttle, are joined in the commercial launch business by a European consortium that has already orbited satellites for U.S. customers via the Ariane rocket. China and Japan also are entering the market with vehicles of their own.

Technologies that are relatively new in the United States have taken little time finding homes around the world. Paging and cellular mobile phone systems are common in Europe and in commercial centers the world over. Even in Moscow, emergency medical and media personnel use a paging system originally built and installed by a British manufacturer for the 1976 Olympics.

There is also the remarkably popular VCR. Although the United States has the world's highest VCR penetration, in Europe VCRs and video rental stores are common, with VCR ownership levels in the United Kingdom and West Germany expected to approach levels in the United States. In the Third World's traditional markets and bazaars pirated copies of American and European television shows and movies are common merchandise. In Venezuela, Brazil, and other Latin American countries, the black-market, illegally imported VCR has become commonplace in middle- and upper-income households despite import bans.[2] In Asia, the Middle East, and Africa VCRs are found, even in lower-income homes, to an extent that government officials sometimes find surprising and, in some cases, ominous. Political speeches on audio- and videocassette, smuggled into Iran and played secretly in cities and villages, have been credited with helping to inspire and organize the successful 1979 Iranian revolution.[3] Worldwide government attempts to prevent the import of VCRs and foreign cassettes, for economic or cultural reasons, generally have been only partially successful.

Promise and Paradox

Jacques Ellul, French sociologist and long-time observer of the social impact of technology, has argued that all technological progress is fundamentally paradoxical in nature.[4] That is, for every problem a technological innovation solves, it creates another. The new electronic media are no exception to Ellul's observation. Although the innovations and trends discussed in this book have positive, even revolutionary, aspects, they also present important new problems of increasing complexity.

Satellite communication provides one of the most dramatic examples of the paradoxical nature of the new electronic media. In only three decades geostationary satellite technology has revolutionized the world of communications 22,300 miles below. Voice, video, and data communications are now possible between virtually any two points on earth, even moving points like ships at sea or aircraft in flight.

Freed by satellites from their terrestrial communication channels, mass media now have virtually unbounded reach. This is possible because satellites have revolutionized not just the technology but the economics of transmission. Distance traveled and number of receivers are no longer the determining factors in the cost of a communication link. Millions of homes can be reached by satellite for the same price as a single home; hundreds of affiliates for the same cost as one. The cost to a local TV news operation of transmitting a live news report from Seattle to Miami is the same as a report from Atlanta to Miami. Consequently we are seeing national and international reach or coverage within the capabilities of what were once local media, and global reach in the case of national media.

The satellite's remarkable ability to extend reach also applies in other communication services. As we have seen, local paging and cellular telephone systems are being linked into national and international systems via satellite; and the national teleconference and many other kinds of electronic meetings have been made possible primarily by the satellite's ability to span great distances with video at bargain prices. Perhaps more than any other technology included in this book, communication satellites provide the infrastructure on which much of the information age is being constructed.

However, true to Ellul's maxim, satellite communication provides revolutionary benefits, but it simultaneously presents difficult new problems on several levels. Perhaps the greatest involves the growing recognition that the benefits of this technology can be derived only through access to the limited number of spaces available to park geostationary satellites. While satellites have dramatically expanded our international communication capabilities, they have created the world's newest controversial natural resource, the GSO or Clarke Belt, which requires international regulation. Debate continues over the most equitable method of assigning parking slots to industrialized nations with active space programs and reserving slots for countries not yet able to launch their own satellites.

The security of satellites provides another paradox. The successful sabotage by "Captain Midnight" of a primary commercial satellite in the United States (see Chapter 5) illustrates the enormous vulnerability of current satellite-based communication systems to intentional interference or disruption. Not only are satellites sitting ducks for earth-bound jamming devices, but a jammer or eavesdropper can operate from virtually anywhere within the satellite's footprint, potentially as much as one-third of the globe. In other words, while conventional geostationary satellites are enormously powerful communication tools, they are at the same time quite vulnerable, more so than terrestrial microwave systems.

Closer to the surface, satellites are challenging the traditional television broadcasting systems around the world in ways no one fully anticipated twenty years ago. In the United States satellites have broken the network's dominance of program distribution to affiliates, given affiliates more autonomy in national and foreign news coverage, and helped make the cable industry a competitor for network audience share. It was the satellite that turned an unremarkable Atlanta television station into the United State's first national station, WTBS, and also gave birth to modern cable programming. Satellite-fed programming transformed the cable industry from a sleepy relayer of over-the-air signals to a genuine competitor for national audience share and programming with the three traditional broadcasting networks.

Thus, while satellite technology has given local stations more autonomy, it also threatens them because it weakens the networks to which they belong. This in turn indirectly threatens local programming on those stations, one of the foundations of the U.S. system of broadcasting. In other countries satellite program distribution competes with the established state broadcasting systems. Furthermore, increasing satellite program distribution across national boundaries and the advent of DBS have increased concerns of some nations that losing control of their airwaves may cause the loss of their cultural identity. Beyond cultural considerations, there is the fear that remote sensing satellites and transborder data flow via satellite may well threaten the sovereignty and national security of less developed nations.

Some critics have suggested that communication satellites, computers, and other new media are being used by the United States and other capitalist countries to extend their economic and political hegemony over lesser developed countries—that is, as weapons of ideological supremacy.[5]

As satellite technology demonstrates, the proliferation of new electronic media is paradoxical because it so often involves some sort of economic or social tradeoff. Other examples include the success of cable television in the United States, which has brought a dramatic increase in viewing options but at the same time threatened the vitality of the country's primary sources of free TV, the commercial broadcasting networks. Teleconferencing technology permits an organization's members to meet more often but can be misused, resulting in less-effective meetings. Telecommuting technology permits persons to work without leaving home but can produce loneliness and isolation for some employees and invite management abuses. Mobile phones, paging systems, and voice messaging systems make possible constant contact but also remove the last remaining buffers between an employee and daily job pressures. The development of an ISDN promises to simplify and unify the transmission of data, voice, and video but would also increase our dependence on a single medium for all our telecommunication capabilities.

In summary, the new electronic media, like all technological innovations, are created to solve existing problems. However, those solutions may in turn create new problems that require more ingenuity to solve than it took to create the new technology originally. Our goal, then, should be to maintain a realistic attitude toward the new electronic media, making the most of their positive attributes, while appreciating their potential for unexpected, perhaps, undesired effects.

Notes

1. See, for example, Teri Robinson, "Telcos Pushing Cable TV Barrier," *MIS Week* (17 October 1988): 1.
2. See Armand Mattelart and Hector Schmucler, *Communication and Information Technologies: Freedom of Choice for Latin America?* (Norwood, N.J.: Ablex, 1985), pp. 33–37.
3. See Douglas A. Boyd and Joseph Straubhaar, "Developmental Impact of the Home Video Recorder on Third World Countries," *Journal of Broadcasting and Electronic Media* (Winter 1985): 5–21.
4. Jacques Ellul, "The Technological Order," *Technology and Culture* (Fall 1962): 394.
5. See Herbert I. Schiller, *Information and the Crisis Economy* (New York: Oxford University Press, 1986).

GLOSSARY

ATV Advanced Television System; any of several new TV technologies designed to enhance the resolution and other capabilities of present-day television.

ACTV Advanced Compatible Television System; an ATV system with a signal capable of being received on present-day television receivers.

AD HOC SYSTEM A temporary teleconferencing system set up for a special event or purpose; not a permanent teleconferencing facility.

ADP Automatic (or Advanced) data processing.

ANALOG A means of encoding or reproducing information using a continuously varying electronic signal. See DIGITAL.

ANTIOPE The French videotext/teletext system; an acronym that roughly translates as "numerical acquisition and televisualization of images organized in pages of characters."

AUDIO CONFERENCING Teleconferencing using voice only; may include use of microphones and loudspeakers to augment telephone handsets.

BANDWIDTH Given in frequencies, the capacity of an electronic transmission system. Usually, the more bandwidth a system requires, the more expensive it is to transmit. TV is a "broadband" system, a telephone voice circuit is "narrowband."

BAUD A measure of the speed of transmission of data; number of elements transmitted per second.

BINARY A coding or counting system with only two symbols or conditions, such as "on/off" or "zero/one"; the format for storing information in computers.

BIT A shortened form of the term "binary digit," the smallest unit of information that can be stored in a computer in binary form.

BITE In a computer's memory, a group of bits (usually eight) that together store a piece of information such as a letter of the alphabet.

BRIDGE In teleconferencing, a device used to interconnect three or more phone lines in different locations.

BURST TRANSMISSION The transmission of data over a circuit at intervals that last only a fraction of a second rather than occupying the circuit continuously.

BYTE See BITE.

CAPTAIN Character and Pattern Telephone Access Information Network; the Japanese videotext system.

CARRIER FREQUENCY The frequency of an electronic signal used to transmit information. See CARRIER WAVE.

CARRIER WAVE An electronic signal, such as that of a television station, onto which information is imposed by varying the signal.

CARS Community antenna relay service; microwave relay systems used by cable operators to transmit cable signals.

CATV Community antenna television; general term for cable TV.

C BAND Part of the electromagnetic spectrum used for terrestrial and satellite microwave transmission, ranging from 4gHz to 6gHz.

CCIR A committee of the ITU which negotiates and recommends technical standards for radio communications and broadcasting.

CCTV Closed circuit television.

CEEFAX Teletext system of the British Broadcasting Corporation (BBC).

CELLULAR MOBILE A mobile radio-telephone service for automobiles and other applications that divides the service area into a number of cells, each with its own transmitting and receiving antenna (Chapter 18).

CHIP A single integrated circuit etched onto a small chip or square made from silicon and other materials.

CIRCUIT A communication link permitting transmission and reception in both directions.

COAXIAL CABLE The cable used in cable TV systems, consisting of several strands surrounded by a sheath of insulation.

CODEC Coder/decoder; a device that digitally encodes and decodes video, audio, or computer data signals for transmission and reception.

COMMON CARRIER A company providing telecommunication services to the public under rates and regulations set by the FCC; does not originate or control the content of the messages it transmits.

COMMUNICATIONS MIX A particular combination of communication media, services, or techniques.

COMPRESSED DIGITAL VIDEO A method of reducing the bandwidth needed to transmit a video signal; only the parts of each frame that are different from the preceding one are transmitted.

COMPUNICATIONS General term for the combination of computers with communication and information technology.

COMSAT Communication Satellite Corporation; government-chartered company that provides satellite communication services and represents the United States in Intelsat.

CPU Central processing unit; the "brains" of a computer where data processing takes place.

CROSS POLARIZATION A technique for doubling the number of TV signals or other transmissions a satellite can carry.

CRT Cathode ray tube; a television screen. May be combined with a receiver in a TV set or may be used as a display device on a computer or data terminal.

C-SPAN Cable Satellite Public Affairs Network; provides coverage of U.S. House of Representatives.

DATABASE A body of information that has been stored in computerized form and can be accessed by computer.

DBS Direct broadcast satellite; a television broadcasting system that beams programming directly from a satellite to small rooftop antennas on subscribers' homes. See Chapter 5.

DEDICATED SYSTEM A communication or information system designed for one particular purpose, such as a dedicated word processing system or a dedicated telephone circuit used only for teleconferencing.

DIGITAL Representing or encoding information by means of separate, discrete measurements or bits of data. See ANALOG.

DIGITAL TRANSMISSION The transmission of sound, images, or data in the form of binary data or bits, rather than as modulations on a carrier wave.

DISH A parabolic or dish-shaped antenna used for sending or receiving signals to communication satellites.

DIVESTITURE The 1984 reorganization of AT&T that created seven independent regional telephone companies. See RBHC.

DOMSAT Domestic communication satellite.

DOWNLINK An earth station that receives transmissions from a communication satellite.

DOWNTIME Period of time when equipment is not working or in service.

EARTH STATION The antenna and other equipment needed on the ground to transmit or receive communication satellite signals; may be a downlink, an uplink, or both.

ECOM Electronic computer-originated mail; system belonging to U.S. Postal Service.

EDP Electronic data processing.

EFT Electronic funds transfer; banking and financial systems that use computers to complete transactions electronically rather than by transfer of paper documents.

ENHANCED SERVICES A category of telecommunication services established by the FCC in Computer Inquiry II in which the transmitted information is significantly changed or restructured by computers, not merely routed by them; not regulated as common carriers.

ELECTRONIC MAIL The forwarding, storage, and retrieval of messages by electronic transmission systems, mostly using digital transmission techniques.

ELECTRONIC OFFICE An "office of the future" in which all information creation, storage, and retrieval and all communication systems are integrated into an efficient electronic network.

EMULATOR A device that makes one piece of telecommunication or computer equipment perform like another one, thus helping to interconnect different types of equipment.

ERGONOMICS The study of the relationship between a machine, such as a computer, and its human users.

FACSIMILE Electronic transmission of printed materials.

FAX Facsimile transmission; a system that transmits printed pictures or documents electronically.

FCC Federal Communications Commission; five-member independent regulatory agency established in 1934 to regulate U.S. telecommunications, including broadcasting.

FIBER OPTICS The technology of using light-transmitting fibers to transmit information. See OPTICAL FIBERS.

FILTER In telecommunications, a device on a transmission line that allows some signals to pass while stopping or trapping others.

FLOPPY DISK A flexible plastic disk coated with material similar to that on recording tape, used to store information in small computers.

FOOTPRINT The portion of the earth's surface covered by the signal from a communication satellite.

FRAME In video technology, a single, motionless, screen-full of video information. Frames transmitted sequentially at a rate of 30 frames per second in North America and Japan and 25 frames per second in Europe create the apparent motion on television screens.

FRAME GRABBER A device that can identify, store and display a single frame of a television picture out of a television signal.

FRANCHISE In cable, the legal document or contract that grants the operator the right to install a cable system and to charge customers for its use.

FREEZE FRAME Also called "slow scan," a type of television transmission used in teleconferencing that permits the display of still pictures only and requires several seconds for each image to be sent. Much more economical than "full motion" video.

FREQUENCY The number of cycles per second (cps) of an electromagnetic transmission, usually described in hertz (kilohertz—1,000 cps, megahertz—a million cps, gigahertz—a billion cps). Generally, high-frequency transmissions can carry more information at greater speeds than low-frequency transmissions.

FULL-MOTION VIDEO In videoconferencing, the use of video links that provide normal, uninterrupted live pictures. See FREEZE FRAME.

GALAXY Name for communication satellites owned by Hughes Aircraft Corporation.

GHZ Gigahertz. See FREQUENCY.

GENERATION A stage in development, such as a "third-generation computer"; In copying and duplication, the number of copies removed from the original item, such as a second-generation tape (a copy made from a copy).

GEOSTATIONARY ORBIT A geosynchronous orbit directly over the equator, 22,300 miles above the surface of the earth. In such an orbit a communication satellite appears to hang motionless in the sky because it is moving at exactly the same speed as the earth is rotating.

GEOSYNCHRONOUS ORBIT An orbit in which a satellite orbits the earth once every twenty-four hours.

GLOBAL VILLAGE A term referring to the potential of telecommunications technology for connecting every spot on the earth with any other, theoretically linking all people as closely together as they sometimes are in small towns; coined by Marshall McLuhan.

HARDWARE In telecommunication and computer equipment, the machinery—the mechanical and electronic parts. See SOFTWARE.

HARDWIRED Any electronic equipment with a permanently installed program designed to carry out certain functions; its function cannot be changed without replacing the circuitry containing the program.

HDTV High-definition television, also "high-resolution television"; a Japanese-developed television system featuring a TV screen with more than twice the lines of the present U.S. system. See Chapter 11.

HEADEND In a local cable television system, the control center where television signals are processed and inserted into the cable distribution system.

HELICAL SCANNING In a videocassette recorder, the manner in which the spinning head reads or records information on the tape, passing across the tape surface at an angle.

HERTZ "Cycles per second," a measurement of the frequency of an electromagnetic transmission or signal. See FREQUENCY.

HOLOGRAPHY A method of storing and projecting a three-dimensional image using lasers.

HYBRID INTERFACE Connection of analog with digital equipment.

INTELLIGENT TERMINAL A computer or communications terminal with the ability to process data on its own, without sending that data to a main computer.

INTELSAT International Telecommunications Satellite Consortium; a group of more than 100 nations organized in 1964 to cooperate in the operation of communication satellites. See COMSAT.

INTERFACE The point at which two systems or pieces of equipment are connected.

INTEGRATED CIRCUIT See CHIP.

INTERSATELLITE LINK A message transmission circuit between two communications satellites, as opposed to a circuit between a single satellite and the earth.

ISDN Integrated Services Digital Network, a communication network under development which allows voice, video and data to be transmitted digitally over a single system at the same time.

ITFS Instructional television fixed service, a band of microwave frequencies set aside for use by public educational institutions such as schools and state educational agencies.

ITU The International Telecommunication Union, an agency of the United Nations that establishes standards and allocates frequencies for telecommunications worldwide.

KHZ Kilohertz. See FREQUENCY.

KU BAND Frequencies from 12gHz to 14gHz used for satellite transmission.

LAN Local area network; equipment used to link together various kinds of electronic office equipment, including computers, word processors, and facsimile machines.

LASER Light amplification by stimulated emission of radiation; a device that emits an extremely narrow beam of pure, high-energy light.

LCD Liquid crystal display; a means of creating numbers and letters using crystals that can be changed from transparent to opaque with an electrical current.

LED Light-emitting diode; a semiconductor device that glows when electric current passes through it, often used for alphanumeric displays on calculators, watches, and other electronic equipment.

LO Local origination; in a cable system, a channel set aside for programs originating from the cable studio rather than from satellite or local broadcast sources.

LOOK ANGLE On a satellite earth station, the angle between the horizon and the antenna; the amount the antenna is tilted up to "see" the satellite.

LPTV Low power television; a secondary television broadcasting service using transmitters with effective coverage of no more than a few miles. See Chapter 5.

MAINFRAME The largest category of computers: mainframes are often room-size, minicomputers are closet-size, and micros are desktop-size or smaller.

MATV Master-antenna television; an antenna system built into a multiple-unit dwelling that allows occupants to receive over-the-air broadcast signals without having to put up their own antennas.

MB Megabytes; a million bytes. See BITE.

MDS Multipoint distribution service; pay television service using a microwave system to transmit movies and other premium programming to subscribers with microwave receiving antennas. See Chapter 10.

MHZ Megahertz. See FREQUENCY.

MICROCIRCUIT Another term for integrated circuit. See CHIP.

MICROWAVE High-frequency radio waves used for telecommunications transmission, usually those above 890 MHz.

MMDS Multichannel Multipoint Distribution Service; an MDS system using two or more channels to transmit multiple program services simultaneously to subscribers.

MODEM Modulator/demodulator; a device which converts digital signals to analog and analog to digital, allowing computer data to be transmitted on telephone lines.

MODULAR Built out of components that can be removed easily and replaced with minimum technical expertise, such as a modular television receiver or a modular telephone receiver.

MODULATION Modifying an electromagnetic transmission in a way that allows it to carry information. AM radio uses amplitude modulation to transmit voice and music; FM radio uses frequency modulation.

MODULE See MODULAR.

MSO Multiple system operator; in the cable industry, a company that operates two or more cable systems.

MTS Mobile Telephone Service; telephone service using radio transmission to link mobile subscribers with the local telephone network. See CELLULAR MOBILE.

MULTIPLEXING Using a transmission channel to carry two or more signals at the same time.

NAB National Association of Broadcasters.

NABTS North American Broadcast Teletext Standard; a high-resolution teletext system backed by the three television networks and AT&T, among others. Not compatible with a competing system called World System Teletext. See WST.

NAPLPS North American Presentation Level Protocol Syntax; the "de facto" U.S. videotext standard, developed by AT&T.

NARROWCASTING Programming an electronic medium with materials aimed at a specific audience, rather than a mass audience; opposite of broadcasting.

NCTA National Cable Television Association.

NEXIS A database offering complete text of stories from major newspapers and wire services.

NODE A point in a communication network where lines of communication come together.

NTSC National Television System Committee; In the United States, a television industry group that develops standards for television broadcasting and receiving equipment. The U.S. TV system is called the NTSC system. See PAL and SECAM.

O&O Owned and operated; a television station that is owned and operated by one of the three major television networks.

OFF LINE Not directly connected with the data processing part of a computer, the central processor; not requiring interaction with the central processor.

OFS Operations fixed service; microwave frequencies set aside for private business and industrial television and other use.

ON LINE Connected to the central processor of a computer; requiring such interconnection.

OPTICAL FIBERS Strands of glass used to transmit light beams originating from LASERs or LEDs. In telecommunications the light is used to carry information, such as telephone conversations or computer data.

ORACLE Teletext system of Britain's Independent Broadcasting Authority.

PABX Private automated branch exchange; a telephone switchboard located on the premises of a multitelephone user. Can also be manually operated (PBX).

PACKET SWITCHING A way of transmitting data or other information through a communication network in which each message is broken down into small digital units, called packets, each with its own identification code or address. Each packet can then take the most efficient route through the system and the message reassembled by computer at its destination using the address codes.

PAGING Also called radiopaging; technology that allows a caller to use the local telephone system in order to contact subscribers carrying pocket receivers that "beep" or otherwise alert them to place a call or receive a message.

PAL Phase alternate line; a system for color television broadcasting used in Europe, not compatible with the U.S. system; see NTSC and SECAM.

PAY-PER-VIEW A pay TV system that allows subscribers to select and pay for individual programs on a one-time basis rather than purchasing a channel for a month or more.

PAY TV A system in which viewers pay by the month for channels of programming, usually movies.

PBX See PABX.

POLARIZATION The direction in which an electromagnetic transmission, such as a TV station signal, vibrates. Determines the orientation of the receiving antenna. See CROSS POLARIZATION.

PPV See PAY-PER-VIEW.

PRESTEL The British videotext system.

PTT Postal, Telegraph and Telephone; refers to government agencies, such as those in Western Europe, which regulate and/or provide centralized telecommunication services, often as state monopolies.

QUBE The interactive cable system developed by Warner Amex in Columbus, Ohio, and now used in its newer franchises.

RAM Random access memory; the part of a computer's memory that can be used to store temporary information, such as the program being used by the operator to accomplish a particular task.

RBHC Regional Bell Holding Company; the seven regional phone companies formed from the twenty-two original AT&T-owned telephone operating companies as a result of the 1984 divestiture.

REAL TIME A telecommunication or data processing system that responds immediately, in the same time frame as the user provides input.

RELAY A switch.

RESOLUTION The amount of detail that can be seen in an image. The resolution of a TV screen is stated in the number of horizontal lines of picture elements the screen displays.

RFP Request for proposals; in cable the document published by a city that announces its desire to receive proposals for a cable system.

ROM Read-only memory; the part of a computer's memory that contains permanent information needed for internal operations. Not available to the user.

SATCOM The series of communication satellites built by RCA.

SATELLITE CARRIER A company that provides communication satellite facilities at rates regulated by the FCC. See COMMON CARRIER.

SBS Satellite Business Systems; first company to launch a communication satellite dedicated solely to business data communications.

SCA Subsidiary communications authorization; permission granted by the FCC to an FM radio station to use certain parts of the FM broadcast signal to transmit material in addition to the regular FM broadcast. See SUBCARRIER.

SCC Specialized common carrier; a company other than the telephone company which offers telephone communication services between cities.

SCRAMBLER In cable and satellite television transmission, a device that alters a program signal electronically so that it can be seen only by persons who have paid for proper decoding devices.

SECAM Sequential Couleur à Mémoire; the color television system used in France, the Soviet Union, and Eastern European countries. Incompatible with the world's other two systems. See NTSC and PAL.

SLOW SCAN See FREEZE FRAME.

SMART TERMINAL A terminal used with a computer or telecommunications system that has some built-in data processing capabilities but not to the extent they may be found in an INTELLIGENT TERMINAL (see).

SMATV Satellite master antenna television; private cable systems installed in multiple-unit dwellings like apartment complexes. See Chapter 10.

SMPTE Society of Motion Picture and Television Engineers.

SOFTWARE The computer programs used to operate computers or computerized telecommunication equipment. See HARDWARE.

SPECTRUM The complete range of electromagnetic frequencies available for all types of telecommunications transmission.

STV Subscription television; pay television services provided by broadcasting movies and other premium programs on a regular television station and scrambling the signal so it can be received only by subscribers with decoders. See Chapter 10.

SRS Subscriber response services; interactive cable television services such as home shopping or classified ads.

STC Satellite Television Corporation; a subsidiary of COMSAT that has received FCC approval for a three-channel direct broadcast satellite service. See DBS.

SUBCARRIER A portion of the broadcast signal of an FM radio station that can be used to transmit programming not available to the general public on normal FM radios. Many operators use it to provide background music service to commercial subscribers equipped with special FM subcarrier receivers.

SUPERSTATION A television broadcasting station whose signal reaches homes across the nation on cable by satellite.

TELECOMMUTING The use of telecommunication technology to allow information workers to work at home, connected to their employers electronically. See Chapter 17.

TELECONFERENCING Electronic meetings; using telecommunication technology to hold meetings or bring individuals into gatherings while remaining in different locations. See Chapter 16.

TELIDON Canadian videotext/teletext system.

TELEFAX Electronic interconnection of facsimile machines for transmission of printed materials. See FAX.

TELEMATIC Americanization of the French "telematique," referring to the combination of computer and telecommunication technologies.

TELEPROCESSING Data processing involving computers in different locations linked by telecommunication equipment.

TELETEXT The one-way transmission of text and graphic information to TV sets using part of the normal television broadcast signal. See Chapter 14.

TELEX Western Union's worldwide commercial telegraphic communication service.

TERMINAL Any device used for sending or receiving information over a communication channel.

TRANSLATOR A low-powered, remote television broadcasting station that picks up the weak signal of a distant full-power station, changes it to a different channel, amplifies it, and rebroadcasts it to nearby viewers. Used to extend signal coverage. See LPTV.

TRANSPONDER A device that receives and retransmits an electromagnetic signal. Present communication satellites have twenty-four transponders, each of which can handle a single color television signal.

TURNKEY SYSTEM A telecommunication system completely installed by one contractor who takes full responsibility from start to finish. All the purchaser has to do is "turn the key," that is, turn the system on and begin operation.

TVRO Television receive-only; a satellite antenna used only to receive television signals, that does not require an FCC license for ownership or construction.

TX See TELEX.

UHF Ultra high frequency; in television broadcasting, the stations operating on channels above channel thirteen.

UNESCO United Nations Educational and Scientific Organization.

UPLINK A satellite earth station that transmits a signal up to a communication satellite rather than receiving it. See DOWNLINK.

VAC Value-added carrier; a telecommunications common carrier that offers additional services beyond merely transmitting information from one to another. VACs normally lease the communication networks they use and install computer equipment that can offer special communication services.

VBI Vertical blanking interval; the portion of a standard television broadcast signal, normally unseen, that is used to transmit text in a teletext system.

VCR Video cassette recorder.

VDT Video Display Terminal; a television monitor used to display computer video output.

VHF Very high frequency; in television, stations broadcasting on channels two through thirteen.

VIDEO CONFERENCING Teleconferencing that uses video links as well as audio. See FREEZE FRAME and FULL MOTION VIDEO.

VIDEODISC A device that can store and play back video and other information on discs similar to phonograph records. See Chapter 15.

VIDEOTEX A generic term referring to all television text information service, including both teletext and videotext.

VIDEOTEXT Two-way home video information services in which the home TV set or terminal is linked to a computer by telephone wire or cable. See Chapter 14.

VIEWDATA A British-originated term for videotext.

VOICE MESSAGING SYSTEM Technology permitting callers to leave or hear recorded voice messages on a computer connected to a phone system.

VSF Voice store and forward; equipment that can store verbal telephone messages digitally and play them back to the receiver at a later time.

VTR Video tape recorder.

WARC World Administrative Radio Conference; international meetings held to negotiate the use of the electromagnetic spectrum and establish ground rules for international telecommunications cooperation.

WESTAR Series of communications satellites belonging to Western Union.

WIRELESS CABLE Another term for MMDS.

WORD PROCESSING The creation, editing, and printing of text using computers or typewriters with internal memory and storage capabilities.

WST World System Teletext; a teletext system being marketed in the United States that uses British CEEFAX/ORACLE technology. See NABTS.

INDEX

ABOUT THE AUTHOR

Loy Singleton received a doctorate in mass communication research from the University of Texas at Austin and taught for nine years at the University of North Carolina at Chapel Hill. He is currently chairman of the Broadcast and Film Communication Department, College of Communication, University of Alabama. His teaching and research interests include electronic media law and regulation, and new telecommunication technologies. His research has appeared in *The American Journal of Policy Analysis and Management*, the *Journal of Communication*, and the *Journal of Broadcasting and Electronic Media*.